ORGANISATIONAL RESILIENCE
Concepts, Integration and Practice

ORGANISATIONAL RESILIENCE

Concepts, Integration and Practice

Edited by
Ran Bhamra

CRC Press
Taylor & Francis Group
Boca Raton London New York

CRC Press is an imprint of the
Taylor & Francis Group, an **informa** business

CRC Press
Taylor & Francis Group
6000 Broken Sound Parkway NW, Suite 300
Boca Raton, FL 33487-2742

First issued in paperback 2019

© 2016 by Taylor & Francis Group, LLC
CRC Press is an imprint of Taylor & Francis Group, an Informa business

No claim to original U.S. Government works

ISBN-13: 978-1-4822-3356-8 (hbk)
ISBN-13: 978-0-367-37739-7 (pbk)

Library of Congress Cataloging-in-Publication Data

Organisational resilience : concepts, integration and practice / editor, Ran Bhamra.
    pages cm
  Includes bibliographical references and index.
  ISBN 978-1-4822-3356-8
  1. Organizational change. 2. Organizational resilience. 3. Crisis management. I. Bhamra, Ran, 1965-

  HD58.8.O695 2016
  658.4'06--dc23
                  2015016934

Visit the Taylor & Francis Web site at
http://www.taylorandfrancis.com

and the CRC Press Web site at
http://www.crcpress.com

For my parents and my daughters

Contents

SECTION I Organisational Resilience Concepts

SECTION II Integration

SECTION III   Practice

Acknowledgements

As a sign of appreciation, the editor wishes to thank all those who have contributed to this book. Firstly, I thank each of the contributing authors who agreed to allow their work to be used in this book, especially all those who readily revised their papers for better fit. Secondly, I would like to thank both Hayley Ruggieri and Cindy Carelli at Taylor & Francis/CRC Press for patiently guiding and supporting this venture, which I am sure can be highly recommended to a wide readership.

Contributors

Emmanuel D. Adamides is an associate professor of operations management at the School of Engineering at the University of Patras, Greece. He holds a BSc from the University of Sussex, UK, an MSc (research) from the University of Manchester, UK and a DrEng from the Democritus University of Thrace, Greece. Prior to joining the University of Patras, he held academic positions in Greece and Switzerland and he was a manager in and consultant to industry. His main research interests are in the application of systems approaches in operations strategy, supply chain management and large-scale innovation and technology management.

Ran S. Bhamra is presently a senior lecturer in the Wolfson School of Mechanical and Manufacturing Engineering at Loughborough University, UK. He is a chartered engineer (CEng) and a member of the Institution of Engineering & Technology (MIET). He has a PhD in resource-based strategy from Cranfield University. Dr. Bhamra joined the school in 2008, prior to which much of his career was spent in a variety of industrial organisations and consultancies that include process improvement, project and operations management. Key research interests include organisational resilience, competence and resource-based theory, sustainable manufacturing strategy and also product–service operations. Dr. Bhamra is a visiting associate professor in the Department of Industrial and Manufacturing Systems Engineering at the University of Hong Kong.

Kevin Burnard is a lecturer in operations management at the University of the West of Scotland. Graduating from Loughborough University with a PhD in organisational resilience and a BEng in manufacturing engineering and management, Dr. Burnard's research and teaching interests focus primarily on organisational strategy and supply chain management. Working within the areas of operations and strategic management, his interest in how organisations are able to perform and function during periods of adversity has led to current research activities related to the concept of resilience.

Amanda Comoretto is a psychologist with 15 years of experience researching processes of resilience development. She earned her PhD investigating a cohort of humanitarian aid workers in a longitudinal study exploring their clinical and occupational outcomes pre- and post-field deployment. Previous to that she worked as a clinical psychologist in Spain and Italy and as a research assistant in the United States, collaborating on a longitudinal study in which a large group of high-risk individuals were followed into adulthood. She is currently active as an international consultant on the topic of stress management and resilience development for companies and academic organisations.

Samir S. Dani joined the University of Huddersfield Business School in February 2014 as professor of logistics and supply chain management. Prior to joining academia, Samir's mechanical engineering background led him to work in the Indian automotive industry for five years. He completed his PhD at the Wolfson School of Mechanical and Manufacturing, Loughborough University, exploring buyer–supplier relationships within a psychological perspective. In 2005, he joined the Wolfson School as lecturer in manufacturing organisation and in September 2008 left to join the School of Business and Economics, Loughborough University as senior lecturer in operations management. Samir has experience leading and working on various projects in a diverse range of management issues from supply chain alignment to leadership to e-commerce. He also has experience teaching a wide variety of management topics at UG, PG, MBA and Exec education levels. He has been an external examiner at the University of Hull, University of Liverpool and Cardiff University.

Uta Hassler heads the Institute for Historic Building Research and Conservation at the Swiss Federal Institute of Technology in Zürich (ETH). Her research interests are in the long-term conservation of values in buildings and building stocks, historic construction techniques and research methods in these fields. Professor Hassler holds a doctoral degree in engineering and a diploma in architecture from the Karlsruhe Institute of Technology. She joined civil service as a member of staff at the Ministry of Finance and completed the leadership academy of the German state of Baden-Wuerttemberg. From 1991 to 2006 she was a professor at Dortmund University, where she established the chair for architecture research, conservation and design. To date, Uta Hassler serves on various commissions (on municipal, state, federal and EU levels) charged with questions of the built environment, academic curricula and research in Germany as well as Switzerland.

Tracy Hatton graduated from the University of Canterbury with a Bachelor of Arts in geography and political science in 1993. After working as a manager in a client service industry in England for many years, she returned to the University of Canterbury to enhance her business skills, graduating with a Master of Business Administration with distinction in 2011. Her MBA project focused on embedding sustainability into both the operations and the teaching within the Canterbury MBA programme. Tracy's PhD research focuses on collaboration as an aid to organisational recovery evaluating how organisations in Canterbury are utilising collaborative approaches to support their recovery and whether these approaches are proving effective.

Michael Henshaw, BSc (Hons), MBA, PhD, MRAeS, MIEEE, MINCOSE, MIEHF leads the Engineering Systems of Systems Research Group. Professor Henshaw worked for 17 years for British Aerospace in aeronautical engineering and development of systems engineering research. He is on the editorial boards of the *Aeronautical Journal* and *SEBoK* (Systems Engineering Body of Knowledge) and has provided advice for parliamentarians on network enabled capability. Some current roles are: NATO Panel in Systems Concepts and Integration, IEEE SMC Society Technical Committee on Systems of Systems, UK MoD Systems of Systems Approach Community Forum.

Ella-Mae Hubbard, MEng, PhD, FHEA, Tech.IEHF is a lecturer in the School of Electronic, Electrical and Systems Engineering at Loughborough University. She has worked in BAE SYSTEMS and in the UK Air Accident Investigation Branch before joining the Engineering Systems of Systems Research Group. Her current interests are in organisational systems, decision making, accident investigation, training and the role of humans in technical systems in general. She is conducting research into pedagogical systems for STEM (science, technology, engineering and mathematics) education.

Hlekiwe Kachali earned a PhD from the University of Canterbury. Dr. Kachali's research focused on the system dynamics of how different sectors within an economy recover from disaster.

Niklaus Kohler is an emeritus professor at Karlsruhe Institute of Technology. Professor Kohler, born in 1941 in Zürich, studied architecture at Georgia Institute of Technology and Ecole Polytechnique Fédérale de Lausanne (EPFL). He was a lecturer at EPFL and invited professor at Ecole Speciale d'Architecture, Paris. He earned his PhD on 'global energy consumption of buildings during their life cycle' in 1985. Professor Kohler was a member of the steering committee of three Swiss technologies transfer programs (Impulsprogramme) in the field of energy and building from 1978 to 1992 and director of the Institute of Industrial Building Production (ifib). Since 2007, Professor Kohler has held teaching and research assignments at ETH Zürich, EPFL Lausanne and the University of Texas at Austin. He is associate editor of *Building Research and Information* and a member of scientific committees in Germany, France, the United Kingdom and China. His research fields are: life cycle analysis of buildings and building stocks and computer supported cooperative planning.

Lee Miles is professor of international relations (IR) at Loughborough University, UK and professor of political science at Karlstad University, Sweden. Prior to this, Lee has held tenured positions at the universities of Liverpool, Hull and Lincoln and has been twice awarded the title of Jean Monnet Professor for his research and teaching on aspects of European Union crisis management. He researches aspects of foreign policy analysis, European crisis management and the politics and foreign policies of the Nordic countries. Indicative work includes *Entrepreneurship in the Polis: Understanding Political Entrepreneurship* (edited with Inga Narbutaité-Aflaki and Evangelia Petridou, Ashgate, 2015) and *Denmark and the European Union* (edited with Anders Wivel, Routledge, 2014). From 2010 to 2013, Lee was also the lead editor of the distinguished IR journal, *Cooperation and Conflict* (Sage).

Philip J. Palin currently serves as the principal investigator and staff consultant on supply chain resilience for the National Academy of Sciences; he consults with the City of Los Angeles and the Federal Emergency Management Agency on disaster logistics and serves as a senior fellow in homeland security with the Graduate School of Rutgers University. Mr. Palin is the principal author for the *Catastrophe*

Preparation and Prevention series from McGraw-Hill. Other publications include *Threat, Vulnerability, Consequence, Risk and Consequence Management*. The *Homeland Security Affairs Journal* has published several pieces by Mr. Palin including Resilience: The Grand Strategy.

Evangelia Petridou is a doctoral candidate in political science at Mid-Sweden University with a background in public administration. Her research interests include theories of the policy process, political entrepreneurship and urban governance. Her substantive policy area interests include economic development, crisis management and territorial cohesion with a special interest in European peripheral areas. She has published in the *Policy Studies Journal* (2014) as well as edited, with Inga Narbutaité-Aflaki and Lee Miles, a volume titled *Entrepreneurship in the Polis: Understanding Political Entrepreneurship* (Ashgate, 2015).

Erica Seville is co-leader of the Resilient Organisations research programme, which, since 2004, has been researching what makes organisations resilient in the face of major crises. Erica has authored over 100 research articles and is regularly invited to speak on the topic of resilience. She is the only non-Australian member to sit on the Resilience Expert Advisory Group, providing advice and support to the Critical Infrastructure Advisory Committee of the Australian Federal Government. Erica has a Bachelor of Civil Engineering Honours degree and a PhD in risk assessment. She is an adjunct senior fellow with the Department of Civil and Natural Resources Engineering at the University of Canterbury and also director of risk strategies research and consulting. Through her company, Risk Strategies Research and Consulting, Erica provides strategic risk management and resilience advice to clients in a variety of sectors including major health providers, the mining and oil and gas industries, construction contractors and critical infrastructure providers.

Carys Siemieniuch, BA, MSc, PhD, MIEHF, Eur.Erg, JP, is a professor in systems engineering and a member of the Engineering Systems of Systems Research Group. A member of INCOSE (International Council on Systems Engineering), she has worked as a systems ergonomist for 24 years in the manufacturing, automotive and aerospace domains. In addition, as a member of the Chartered Institute of Ergonomics and Human Factors, she has extensive and varied consultancy expertise, turning her applied research into practical advice for a number of organisations including the European Parliament, the UK MoD and a range of engineering organisations in the United Kingdom and Europe.

Murray Sinclair, BSc, MSc, EngD, FIEHF, MINCOSE, is a visiting fellow at Loughborough University. He has been a member of the UK Nuclear Safety Advisory Council. He is a systems ergonomist of some 40 years standing and his interests are in the understanding of organisational processes of manufacturing and the management of knowledge. Recently, due to the pervasion of information technology into society and the lives of individuals, his interests now include the assurance of ethical behaviour by autonomous and semi-autonomous systems, including cyber-physical systems and robots.

Joanne R. Stevenson graduated from the University of Illinois with a Bachelor of Science (Hons) in natural resource and environmental science (2007). Joanne worked as a contract researcher at the University of Dundee in Scotland following the 2007 UK floods. She then earned her Master of Science (2010) in geography from the University of South Carolina's Hazards and Vulnerability Research Institute. Her master's thesis focused on understanding spatial and temporal trends in residential reconstruction in Mississippi following Hurricane Katrina. Joanne's PhD research focuses on the recovery of organisations within Central Business Districts (CBD) affected by the September 2010 and February 2011 earthquakes in Canterbury, New Zealand.

Christos Tsinopoulos is a senior lecturer in operations and project management at Durham University Business School and his research explores aspects of supply chain integration, risk and innovation. He has a degree in mechanical engineering from the University of Sheffield and a PhD in engineering from Warwick. He, his PhD students and colleagues explore research questions associated with the mechanisms through which integration with customers and suppliers improve an organisation's ability to innovate. To increase the impact of his research he has been working with the Institution of Mechanical Engineers (IMechE) on their flagship Manufacturing Excellence Awards scheme to provide research led feedback to UK manufacturers on improving their operations.

John Vargo is a senior researcher and co-leader of the Resilient Organisations research programme based at the University of Canterbury. His interests focus on building organisational resilience in the face of systemic insecurity in a complex and interconnected world. John was a practicing certified public accountant (CPA) in the United States prior to moving to New Zealand and has been involved in industry for over 30 years. His research interests are in organisational resilience, information security, risk management, e-commerce and strategic planning. In addition to his academic role, John has filled a range of senior management roles at the University of Canterbury in support of major change initiatives including: dean of commerce, pro vice-chancellor (business and economics), director of ICT, chief operating officer, chief financial officer and assistant vice chancellor for student services.

Jelena V. Vlajic works as a lecturer at Queen's University, Belfast since 2012. She earned a doctoral degree from Wageningen University in the Netherlands. Her PhD project was related to the design of robust food supply chains, vulnerability assessment and disturbance management. Her research interests include supply chain vulnerability and risks, supply chain robustness and resilience, environmentally friendly logistics practices, supply chain (re)design, as well as teaching and research methodologies in operations management and logistics. She has authored and co-authored seven articles published in peer-reviewed international journals. Jelena also works as reviewer for several international journals and book publishers.

Zach Whitman earned his BA with a major concentration in geology and a minor concentration in geography from Colgate University in the United States. He then continued on to the University of Canterbury to study hazard and disaster management, with a specific focus on business resiliency in all hazard environments. His work looks at the effects external aid has upon small- to medium-sized enterprise resilience in both rural and urban settings in New Zealand.

Tom Wilson is a lecturer in hazard and disaster management in the Department of Geological Sciences at the University of Canterbury. His doctoral research analysed the vulnerability of pastoral farming systems to volcanic ashfall hazards. Associated research included developing risk assessment tools and models to assess loss from volcanic hazards using geospatial platforms (GIS) – as part of the Riskscape project. Since beginning as a lecturer at the University of Canterbury, he has developed the Volcanic Ash Testing laboratory (VAT Lab) which investigates the vulnerability of critical infrastructure and primary industries to volcanic ashfall. Tom's other main research interest is understanding the resiliency of rural and isolated communities to natural disasters. This has focused on the impact of recent disasters in the Canterbury region from snowstorm and flooding events and the assessment of natural hazard risk for isolated communities.

Lili Yang is a reader in information systems and emergency management at Loughborough University in the United Kingdom. She is a fellow of the British Computer Society and a chartered IT professional. Her research has been supported by many research funding bodies and industries such as EU, EPSRC, the Technology Strategy Board, British Council, the Royal Society, the Royal Academy of Engineering, UK Fire and Rescue Services, BAE Systems, etc. As principal investigator, she has led a number of projects with the total budget reaching to over £3million. Her publications appear in *Information Systems Research, European Journal of Operational Research, Technological Forecasting and Social Changes,* all top quality journals.

Introduction

A fundamental question that organisations have traditionally asked, and one that a considerable portion of management schools are set up to answering is; how can we, as an organisation, become more competitive? Hitherto, there have been innumerable answers coming from disciplines such as operations management, supply chain management and strategic management to name but three. As we now optimise existing approaches, we have not only become more interconnected as independent entities (organisations) but also progressively more complex – we have begun to realise and sense our 'apparently' increased vulnerability to organisational and business disruption from different sources. The discipline of continuity management emerged many years ago in order to focus on the specifics keeping a business going. However, in reality continuity management is essentially returning a business to 'business as usual', and nothing more. Resilience then should be seen as an overarching concept that not only enables organisations to continue with business as usual, but also to learn, progress and flourish, beyond business as usual. In short, it can be a constructive mechanism to becoming more competitive. However, the underlying key point about resilience as a concept is that it is concerned with survival, pure and simple, but which will likely involve transformation.

With our ever-increasing realisation about the reality of geo-political instability and fragility of the environment, the questions of risk and uncertainty have become paramount. Certainly understanding and dealing with supply chain risk and resilience is very important (food supply chains in particular), but so is economic and financial resilience. This line of thought could of course naturally lead into areas of ethical social behaviour and sustainability, but this is not the main aim of this book. Our resilience to natural hazards, or undeniably our lack thereof, has been a key driver in our development and study of resilience. However, the nature of societal resilience is highly important, and in particular, how an interconnected and complex society can deal with the threat and consequences of natural disasters, terrorism, pandemics, social unrest, cyber-attacks, and also global financial breakdown. In bringing together the chapters contained in this book the editor has tried to capture the diversity of, and to some extent, the depth of current thinking about research on the concepts of organisational-related resilience.

The chapters in this book are organised into three broad sections. Section I captures the fundamental concepts and clarifies some underlying ideas from diverse fields of resilience-related research. Section II looks at how some of these concepts and ideas have been integrated into specific research activity and used to further develop their respective fields of enquiry. The connecting of concepts and ideas to existing ones, in this way, readily helps to progress the development of theory. Section III focuses on aspects of real-world practice and experience.

This book is intended as a broad introduction to rapidly growing interest in the field of resilience-related research, knowledge and practice. This book will be of interest to academics, practitioners, business people and also, as a background or primer, to students who study courses in risk, resilience and its management. Given these intentions, the editor hopes to clarify the commonality of concepts and practice that exists amongst disparate research disciplines and to also establish a brief but singular 'go-to' work for anyone interested in resilience.

Ran Bhamra

Section I

Organisational Resilience Concepts

The aim of this section is to try to both capture and indicate some of the breadth of lenses with which resilience is viewed; indeed the works included here in Chapters 1 through 4, although only a sample, are very representative. Chapter 1 begins with a full update of a recent work by Bhamra et al. on organisational resilience. This chapter is a wide-ranging review of resilience perspectives, concepts and methodologies utilised in research in this burgeoning field. Following investigation into different definitions of core resilience related terms, the authors call for better understanding of whether resilience is a philosophy, a capability, a business measure or a complex facet of organisations. In Chapter 2, Hassler and Kohler, from the perspective of built environment, encompass the systems view as applied to infrastructural resilience. Here, they consider resilience as a set of different capitals, namely – natural, physical, economic, social and cultural. They put forward that each of these capitals have different timescales within which factors such as stability, uncertainty, robustness and so on, will impact. Importantly, the authors assert that resilience thinking must be incorporated into sustainable management of the built environment.

In Chapter 3, Comoretto considers the notion of resilience as applied to human individuals. Here, resilience is about the factors that promote well-being in people who operate in highly stressful environments, in this case humanitarian aid workers. The concepts in this chapter include the consideration of aspects such as cognitive protective factors, which comprise the nature of human motivation and coping skills. This study provides an interesting insight into what we may call resilience within individuals. In Chapter 4, Miles and Petridou explore resilience from the political perspective, in the context of decision making during crisis management. The authors assert that successful resilience is made possible by certain entrepreneurial actions that occur within the 'gaps' that lie between resilient systems and resilience planning. This work is intriguing as 'entrepreneurial/political resilience' may not be an area that readily comes to mind when we first think of resilience, but nonetheless it appears to be a crucial element to understanding the completeness of the resilience concept.

1 Resilience
The Concept, a Literature Review and Future Directions

Ran S. Bhamra, Kevin Burnard and Samir S. Dani

CONTENTS

1.1 INTRODUCTION

In an ever-more interconnected world, countries, communities, organisations and individuals are all subjected to a diverse and ever-changing environment. The threats that this often turbulent environment poses can vary in both severity and frequency. Threats may also originate internally or externally to a system and create far reaching implications and impacts. Often, the impacts of these events are felt across the world. An event in one area can often have disastrous effects in another (Juttner 2005). Events can also take many forms as highlighted by several recently publicised occurences including the 2008 global financial crisis, the 2010 Haiti and Chile earthquakes and the unfolding events of the Fukushima Daiichi nuclear power plant disaster following the 2011 Tohoku earthquake and tsunami. Natural disasters, pandemic disease, terrorist attacks, economic recession, equipment failure and human error can all pose both a potentially unpredictable and severe threat to the continuity of an organisation's operations and infrastructure. The difficulty in proactively addressing these events lies in the fact that disasters are a multifarious concept, composed of many different elements that seem to defy any precise definition (Alexander 2003). It is often only through hindsight that disasters look like the events for which individuals, communities, organisations and countries should have prepared. Additionally, it is not only disasters that pose a threat but also small uncertainties or deviations that can cause significant challenges, particularly within the context of organisations and their associated networks. In order to withstand these challenges, it is essential that sufficient effort be channelled into developing robust and resilient organisational systems capable of withstanding and responding to these uncertainties and challenges.

This chapter explores the literature surrounding the concept of resilience within an organisational context. The term resilience is used in a wide variety of fields that include ecology (Walker et al. 2002), metallurgy (Callister 2003), individual and organisational psychology (Barnett and Pratt 2000, Powley 2009), supply chain management (Sheffi 2005), strategic management (Hamel and Valikangas 2003) and safety engineering (Hollnagel et al. 2006). Although the context of the term may change, across all these fields the concept of resilience is closely related with the ability of an element or system to return to a stable state after a disruption. When the notion of resilience is applied to organisations, this definition does not drastically change. Resilience within an organisational context is related to both individual and organisational responses to turbulence and discontinuities. Following this broad definition, this chapter looks to further develop the understanding of the features and dynamics of resilience to help provide an insight into how organisations may be able to effectively manage uncertainty and overcome disruptive events and threats.

1.2   SUSCEPTIBILITY TO DISASTERS AND DISRUPTIONS

On 21 September 1999, a devastating earthquake measuring 7.3 on the Richter scale hit central Taiwan. The Chi-Chi earthquake was the largest earthquake in the area in almost 65 years, killing 2300 people, injuring 8000 and destroying 100,000 homes (Chen et al. 2007). The impact of the disaster had severe consequences for infrastructure and many organisations. The total industrial production losses were estimated at $1.2 billion following the devastating earthquake (Papadakis 2006). The Hsinchu Industrial Park lay within 70 miles of the earthquake epicentre and included many large-scale semiconductor fabrication facilities that were estimated to account for approximately 10% of the world's production of computer memory chips. Several other computer components were also manufactured within this industrial park. The impact of the disaster on global computer production was dramatic. Component supply was significantly constrained during subsequent months due to the damage caused by the earthquake (Papadakis 2006), and companies such as Apple, Compaq, Dell, Gateway and IBM were all affected by the supply disruption (Papadakis 2006).

In March 2000, lightning struck an electrical cable causing power fluctuations throughout the state of New Mexico. Due to electrical fluctuations, a minor fire developed in a small production cell at a Philips semiconductor plant in Albuquerque

(Norrman and Jansson 2004). Although not a major incident, the 10-min fire had drastic implications for the mobile phone producers Ericsson and Nokia (Tang 2006). The fire took place in one of the plants high-grade clean rooms, and due to the fire, smoke and sprinkler water, the production of components was delayed for 3 weeks. After 6 months, the yield of radio-frequency chips was still only at 50% – it would take years to get new equipment delivered and installed and operationally optimised (Norrman and Jansson 2004).

The deepwater horizon drilling rig provided the platform for one of the largest environmental accidents in decades. On 20 April 2010, a sudden explosion aboard the rig claimed the lives of 11 people and a ruptured pipeline on the seabed released huge volumes of oil into the Gulf of Mexico (Harlow et al. 2011). The National Oceanic and Atmospheric Administration (NOAA) estimated as much as 210,000 gallons of oil leaked through the ruptured pipe into the ocean each day (Muralidharan et al. 2011). The resulting oil spill caused significant environmental damage across the U.S. Gulf Coast. The oil flowed for 3 months until the leak was stopped, a total of 87 days. British Petroleum (BP) has subsequently spent billions of dollars in compensation and clean-up efforts.

The 2011 Tohoku earthquake carried dramatic implications for Japan and much of the surrounding area (Akahane et al. 2012). The estimated 9.0 magnitude earthquake caused large-scale tsunamis off the coast of Fukushima. The resulting tidal surge caused significant damage across the region including extensive damage to the nuclear reactors within the Fukushima Daiichi nuclear power plant. The catastrophic damage caused by the tsunamis resulted in the release of hazardous radioactive materials into the local environment. In the aftermath of the event, the Japanese government set up a tightly controlled restricted area and evacuated residents within a 20 km radius of the impacted area (Akahane et al. 2012).

These examples highlight the dramatic impact that disasters can carry and the major threat they pose to the incumbency and performance of an organisation. In addition to the physical damage and hardship, disruptions can have a direct effect on an organisation's ability to get finished goods into a market and provide critical services to customers. However, how do some organisations overcome these events while others fail? What enables these organisations to adapt and transcend these events? And what sets these organisations apart? Certainly organisations will have business continuity plans (Cerullo and Cerullo 2004) and disaster recovery plans in place; however, unless these procedures and plans can be intuitively applied during fast moving crises, the plans will not be effective (Seville et al. 2006). As a result, a new proactive approach to the management of disruptive and disaster events is required.

To address this need, this chapter presents a critical synthesis of literature towards characterising resilience within an organisational context. While resilience within organisations provides the central focus of this study, ecological, individual, community, socio-ecological and supply chain perspectives are also included. Section 1.3 presents an overview of the methodological considerations of the literature review. The resilience-based literature is then reviewed in its widest perspective. This is followed by a more focused consideration of resilience centred on organisations and includes system perspectives, vulnerability and adaptive capacity.

1.3   REVIEW METHODOLOGY

The objective of this research is to identify the diversity within academic thinking in relation to 'resilience' as a concept, and thereafter, to identify gaps, issues and opportunities for further study and research using this concept. A literature review is an integral part of any research exercise (Easterby-Smith et al. 2002). The review was undertaken in order to gain a holistic view of the diverse and interdisciplinary viewpoints related to the concept. It was felt that this was a necessary step in structuring the literature base to support any future study and research. While organisational resilience provides the central focus of the study, the identified resilience-based literature is reviewed in its widest perspective in order to provide a critical insight into the multifaceted concept. This is then followed by a more focused consideration of resilience centred on organisations.

In conducting the literature review, the process followed the stages outlined by Srivastava (2007) and the reviewed literature is presented in Tables 1.1 through 1.5. Table 1.1 outlines the various perspectives, concepts and methodologies within the reviewed resilience literature, while Table 1.2 outlines several definitions related to resilience. The subsequent tables then quantify the reviewed literature. These tables not only provide an overview of the reviewed literature but also illustrate the range of features and considerations related to resilience within a variety of perspectives.

1.3.1   DEFINING UNIT OF ANALYSIS

A single research article/book: The limitation for this resource was that the article/book was available easily through library databases or Internet databases such as Google Scholar or Science Direct. Following the sampling of the literature base, 100 articles were retained for classification.

1.3.2   CLASSIFICATION CONTEXT

The classification of the reviewed literature is presented in Table 1.1. The classification scheme applied is comprised of three elements: perspectives, topics/concepts and methodologies. The classification stage also involved studying the different definitions of resilience under the perspectives. The dominant perspectives and definitions are presented in Table 1.2.

Identified categories included

Perspectives: ecological, individual, socio-ecological/community, organisational and supply chain
Topics/concepts: behaviour and dynamics, capabilities and strategy and performance
Methodologies: theory building, case study and survey and model/framework

It should be noted that literature related to system and networks were typically classified within the socio-ecological/community perspective. It was felt that this perspective was the most representative of the subject and themes addressed within the articles.

1.3.3 Material Evaluation

The material was analysed and sorted according to the classification context. Following this classification, a narrative review of the retained articles was conducted. This led to the identification of relevant issues and features related to organisational resilience.

1.3.4 Collecting Publications and Delimiting the Field

The literature review focused upon books, edited volumes, journal articles, research reports and conference papers. Library databases and Google Scholar were used with a keyword search using some important keywords and phrases such as 'resilience', 'disaster', 'disruption', 'adaption', 'threat', 'risk' and 'uncertainty'. Overall 200 articles were identified. The articles which provided a direct link to the concept of 'resilience' were considered for the review. Finally, 100 articles were retained and analysed using the classification scheme.

The key aspect of the review was to achieve a wide interdisciplinary perspective for the concept of 'resilience' and to investigate whether there was cross learning between disciplines within the facets of the classification scheme. The overall objective was to identify those concepts which would inform resilience within organisations. Table 1.1 outlines the reviewed literature in relation to the classification scheme.

1.4 WIDER RESILIENCE LITERATURE

Although there has been increasing awareness of the concept within recent years, 'resilience' has received little systematic empirical work and independent attention (Sutcliffe and Vogus 2003). Predominately, resilience-based literature has followed a largely theoretical approach; focusing instead on conceptual development and related areas. This has created a diverse literature base across several disciplines. Following the work of several authors, resilience-based literature can be broadly grouped into three general areas of classification that correlate to the elements of resilience as identified by Ponomarov and Holcomb (2009). These are

- Readiness and preparedness
- Response and adaption
- Recovery or adjustment

As evidenced by Tables 1.1 and 1.2, while several authors attempt to broadly cover all of these general areas within a study, individually each area has received little systematic attention and empirical-based study. In recent years this has started to change. As the field has begun to develop, articles and publications have typically become more focused on a particular context, concept or factor related to resilience. Unsurprisingly, there is a large cross linkage between work related to the ecological perspective of resilience and other contexts.

The term was first popularised by Holling in 1973 within the seminal work titled *'Resilience and Stability of Ecological Systems'*. The work has formed the foundation for most studies of the concept of ecological resilience as well as various other forms of resilience. Holling (1973) outlines how altering views of behaviour within ecological systems can create different approaches to the management of resources. To this end, Holling (1973) presents the viewpoints 'resilience' and 'stability'. These have been further extended to form the terms 'ecological resilience' and 'engineering resilience' (Gunderson 2000, Walker et al. 2002). The 'resilience' viewpoint outlined by Holling (1973) emphasises the domains of attraction and the need for persistence through defining resilience as a measure of systems persistence and the ability to absorb disturbances and still maintain the same relationships between system entities. The 'stability' viewpoint emphasises maintaining the equilibrium within a predictable world and accumulating excess resources with minimum fluctuation of the system and is then defined as the ability of a system to return to an equilibrium state after a disturbance (Holling 1973). Within these viewpoints, resilience is presented as both an active process related to the response of an element or system as well as a feature of adaption and change. Resilience thereby describes the capacity of an element or system to cope under challenging circumstances.

The concept of resilience is both multidisciplinary and multifaceted. The notion of resilience is firmly grounded within ecology and the working definitions used by many authors developed following Holling's (1973) original research relating to ecosystem stability. There have been several definitions proposed for resilience, each slightly altered dependent on context. Table 1.1 gives the diversity of resilience definitions and also highlights some of the distinctions between them.

Table 1.1 outlines the reviewed resilience literature to reveal the research contributions regarding the perspectives taken, concepts discussed and the research methodologies utilised. Although the literature review has considered a classification methodology, the articles reviewed represent only a relative sample of the literature base. However, it is interesting to study this sample in order to provide an insight into which areas are the most strongly represented or acknowledged within resilience-based literature.

From Tables 1.1 and 1.2 it can be seen that the concept of resilience is shown to remain essentially constant regardless of its field of enquiry and has much to inform the fields of organisation theory, strategy and operations management. This analysis identifies a number of areas for advancing resilience research, in particular: the relationship between human and organisational resilience, as well as understanding the interfaces between organisational and infrastructural resilience.

1.4.1 THE PERSPECTIVES

Table 1.3 outlines the various perspectives within the reviewed literature. From the table it is evident that, out of the articles selected for the review, there has been a balanced view across four perspectives ranging from the individual to the organisation, community and the ecology. Supply chain resilience, which perhaps sits between the organisational and community perspectives, has gained more focus post 2001.

TABLE 1.1

Perspectives, Concepts and Methodologies within Resilience Literature

Author(s) (year)	Perspectives					Topics/Concepts				Methodologies			
	Ecological	Individual	Socio-Ecological/ Community	Organisational	Supply Chain	Behaviour and Dynamics	Capabilities	Strategy	Performance	Theory Building	Case Study	Survey	Model/ Framework
Holling (1973)	☐	☐				☐	☐			☐	☐		
Thomson and Lehner (1976)	☐	☐							☐	☐	☐		
Garmezy and Masten (1986)		☐				☐					☐		
Werner (2009)		☐				☐					☐		
Holling (1996)	☐	☐				☐				☐	☐		
Horne and Orr (1998)		☐		☐		☐				☐			
Mallak (1998)		☐		☐		☐				☐			
Masten and Coatsworth (1998)						☐	☐			☐			
Peterson et al. (1998)	☐					☐	☐			☐			☐
Miller and Whitney (1999)			☐			☐	☐			☐	☐	☐	
Adger (2000)			☐			☐		☐		☐			☐
Gunderson (2000)	☐					☐	☐			☐			
Luthar et al. (2000)		☐				☐	☐	☐		☐	☐		
Mitchell et al. (2000)	☐					☐				☐			
Nystrom et al. (2000)	☐			☐									
Paton et al. (2000)	☐		☐			☐		☐	☐	☐	☐		☐
Carpenter et al. (2001)			☐			☐			☐	☐			
Comfort et al. (2001)			☐			☐			☐	☐			

(Continued)

TABLE 1.1 (Continued)
Perspectives, Concepts and Methodologies within Resilience Literature

Author(s) (year)	Model/ Framework	Survey	Case Study	Theory Building	Performance	Strategy	Capabilities	Behaviour and Dynamics	Supply Chain	Organisational	Socio-Ecological/ Community	Individual	Ecological
	Methodologies				Topics/Concepts				Perspectives				
Holling (2001)				□		□	□				□		
Paton and Johnston (2001)				□		□					□		
Coutu (2002)													□
Folke et al. (2002)			□	□			□				□		
Rudolph and Repenning (2002)	□			□				□		□			
Walker et al. (2002)	□							□			□		
Bruneau et al. (2002)	□					□							
Elmqvist et al. (2003)			□	□				□					□
Fiksel (2003)				□	□		□	□					
Hamel and Valikangas (2003)				□		□				□			
Klein et al. (2003)				□			□	□		□	□		
Riolli and Savicki (2003)								□					
Starr et al. (2003)	□			□		□				□			
Sutcliffe and Vogus (2003)	□			□	□			□		□			
Vis et al. (2003)			□	□		□				□	□		
Bodin and Wiman (2004)	□			□				□					□
Bonanno (2004)				□				□				□	

(Continued)

TABLE 1.1 (Continued)
Perspectives, Concepts and Methodologies within Resilience Literature

Author(s) (year)	Ecological	Individual	Socio-Ecological/ Community	Organisational	Supply Chain	Behaviour and Dynamics	Capabilities	Strategy	Performance	Theory Building	Case Study	Survey	Model/ Framework
			Perspectives				*Topics/Concepts*			*Methodologies*			
Christopher and Peck (2004)					□	□			□		□		
Dalziell and McManus (2004)				□			□			□			
Rice and Sheffi (2005)					□	□		□			□		
Walker et al. (2004)	□		□										
Allenby and Fink (2005)			□						□	□			
Cumming et al. (2005)	□					□				□	□		□
Hughes et al. (2005)						□				□			
Juttner (2005)					□							□	
Sheffi (2005)					□			□	□	□			
Bonanno et al. (2006)		□		□		□					□	□	
Fiksel (2006)			□			□				□			□
Gallopin (2006)	□		□			□				□			□
Hawes and Reed (2006)		□				□		□					
Hind et al. (1996)				□			□		□			□	
Hollnagel et al. (2006)				□		□				□			
Luthans et al. (2006)		□		□						□			
Manyena (2006)			□					□					

(Continued)

TABLE 1.1 (Continued)
Perspectives, Concepts and Methodologies within Resilience Literature

Author(s) (year)	Perspectives					Topics/Concepts				Methodologies			
	Ecological	Individual	Socio-Ecological/ Community	Organisational	Supply Chain	Behaviour and Dynamics	Capabilities	Strategy	Performance	Theory Building	Case Study	Survey	Model/ Framework
McDonald (2006)				□		□				□			
Ong et al. (2006)		□				□					□		
Reich (2006)		□				□				□			
Rutter (2006)		□				□				□			
Seville et al. (2006)				□		□			□		□		
Tang (2006)					□			□		□			
Masten and Obradovic (2007)	□	□				□				□			
McManus et al. (2007)				□		□	□			□	□		
Nelson et al. (2007)						□	□	□	□	□	□		□
Sheffi (2007)		□			□	□	□		□	□			
Vogus and Sutcliffe (2007)				□						□	□		
Waters (2007)					□	□	□		□	□			
Youssef and Luthans (2007)		□				□					□	□	
Youssef and Luthans (2007)		□				□	□					□	
Falasca et al. (2008)			□		□	□				□			□
Norris et al. (2008)			□			□				□			
Brand (2009)	□					□				□			□

(Continued)

TABLE 1.1 (*Continued*)

Perspectives, Concepts and Methodologies within Resilience Literature

Author(s) (year)	Perspectives					Topics/Concepts				Methodologies			
	Ecological	Individual	Socio-Ecological/ Community	Organisational	Supply Chain	Behaviour and Dynamics	Capabilities	Strategy	Performance	Theory Building	Case Study	Survey	Model/ Framework
Brand (2009)	✓					✓				✓			
Crichton et al. (2009)				✓		✓				✓			
Petchey and Gaston (2009)	✓					✓		✓		✓	✓		
Ponomarov and Holcomb (2009)					✓	✓	✓						
Powley (2009)		✓				✓		✓		✓	✓		✓
Smith and Fischbacher (2009)				✓		✓			✓	✓			✓
Folke et al. (2010)			✓			✓					✓		
Gibson and Tarrant (2010)				✓		✓	✓			✓			✓
Ruiz-Ballesteros (2011)			✓			✓			✓	✓			✓
Burnard and Bhamra (2011)				✓		✓				✓			
Derissen et al. (2011)			✓			✓				✓			
Engle (2011)			✓						✓			✓	
Goodman et al. (2011)				✓	✓	✓	✓			✓	✓		
Harwood et al. (2011)		✓					✓			✓			
Jüttner and Maklan (2011)						✓							
Lengnick-Hall et al. (2011)				✓						✓			
Steen and Aven (2011)													✓

(Continued)

TABLE 1.1 (Continued)
Perspectives, Concepts and Methodologies within Resilience Literature

Author(s) (year)	Perspectives					Topics/Concepts				Methodologies			
	Ecological	Individual	Socio-Ecological/Community	Organisational	Supply Chain	Behaviour and Dynamics	Capabilities	Strategy	Performance	Theory Building	Case Study	Survey	Model/Framework
Ainuddin and Routray (2012)			□						□			□	□
Henry and Ramirez-Marquez (2012)			□						□	□			□
Kachali et al. (2012)				□			□		□			□	
Poortinga (2012)			□			□						□	
Tveiten et al. (2012)				□			□	□			□		
Wilkinson (2012)			□	□		□				□			
Aleksić et al. (2013)				□		□							□
Johnson et al. (2013)					□	□					□		
Shaw and Maythorne (2013)			□	□				□			□		
Hodbod and Adger (2014)			□			□				□			
Pal et al. (2014)				□		□			□		□	□	
Pereira et al. (2014)					□		□			□			
Scholten et al. (2014)					□		□	□			□		□
Stevenson and Busby (2015)					□			□		□	□		

TABLE 1.2
Definitions of Resilience

Author	Context	Definition/Characterisation
Holling (1973)	Ecological systems	The measure of the persistence of systems and of the ability to absorb change and disturbance and still maintain the same relationships between state variables.
Masten et al. (1990)	Psychology	The process of, capacity for or outcome of successful adaptation despite challenging or threatening circumstances.
Horne and Orr (1998)	Organisational	Resilience is the fundamental quality to respond productively to significant change that disrupts the expected pattern of events without introducing an extended period of regressive behaviour.
Gunderson (2000)	Ecological systems	The magnitude of disturbance that a system can absorb before its structure is redefined by changing the variables and processes that control behaviour.
Paton et al. (2000)	Disaster management	Resilience describes an active process of self-righting, learned resourcefulness and growth. The concept relates to the ability to function at a higher level psychologically given an individual's capabilities and previous experience.
Carpenter et al. (2001)	Socio-ecological systems	The magnitude of disturbance that a system can tolerate before it transitions into a different state that is controlled by a different set of processes.
Coutu (2002)	Individual	Resilient individuals possess three common characteristics. These include an acceptance of reality, a strong belief that life is meaningful and the ability to improvise.
Walker et al. (2002)	Socio-ecological systems	The ability to maintain the functionality of a system when it is perturbed or the ability to maintain the elements required to renew or reorganise if a disturbance alters the structure of function of a system.
Bruneau et al. (2003)	Disaster management	The ability of social units to mitigate hazards, contain the effects of disasters when they occur and carry out recovery activities that minimise social disruption and mitigate the effects of future earthquakes.
Hamel and Valikangas (2003)	Organisational	Resilience refers to the capacity to continuous reconstruction.
Bodin and Wiman (1994)	Physical systems	The speed at which a system returns to equilibrium after displacement, irrespective of oscillations, indicates the elasticity (resilience).
Tilman and Downing (2004)	Ecological systems	The speed at which a system returns to a single equilibrium point following a disruption.

(Continued)

TABLE 1.2 (*Continued*)
Definitions of Resilience

Author	Context	Definition/Characterisation
Walker et al. (2004)	Ecological systems	The capacity of a system to absorb a disturbance and reorganise while undergoing change and while retaining the same function, structure, identity and feedback.
Hollnagel et al. (2006)	Engineering	The ability to sense, recognise, adapt and absorb variations, changes, disturbances, disruptions and surprises.
Luthans et al. (2006)	Psychology	The developable capacity to rebound from adversity.
McDonald (2006)	Organisational	Resilience conveys the properties of being able to adapt to the requirements of the environment and being able to manage the environment's variability.
Linnenluecke and Griffiths (2010)	Organisational	The capacity to absorb impact and recover.
Braes and Brooks (2011)	Organisational	Resiliency relates to a personality characteristic and resilience refers to a dynamic developmental process.

TABLE 1.3
Perspectives in Resilience Literature

Perspectives				
Ecological	Individual	Socio-Ecological/Community	Organisational	Supply Chain
17	19	27	30	15

1.4.2 Topics/Concepts

Considering Table 1.4, it is evident that behaviour and dynamics feature in the majority of the reviewed articles. A total of 68 articles out of the sample of 100 addressed this area. Although there is considerable overlap between topics and concepts within the review classification scheme, it is clear that this area has received the most specific attention within the realm of resilience across all perspectives. This is possibly due to the theoretical and conceptual features of the concept of resilience being developed.

1.4.3 Methodologies

From Table 1.5, it is evident that theory building has been the main focus of researchers within the area of resilience. This again illustrates the nature of this developing concept. Case study and model development have been used roughly in 40% of the reviewed articles. However, the survey has not been a preferred methodology for the study of resilience. Following the evidence of Table 1.1 it can be seen that as the area of resilience-based research has developed, the focus has become increasingly empirically focused.

TABLE 1.4
Topics/Concepts Studied under Resilience

Topics/Concepts			
Behaviour and Dynamics	Capabilities	Strategy	Performance
68	23	21	21

TABLE 1.5
Methodologies Applied within Research

Methodologies			
Theory Building	Case Study	Survey	Model/Framework
67	30	11	23

1.5   ORGANISATIONAL LEVEL AND SYSTEMS RESILIENCE

Fundamentally, the concept of resilience is closely related with the capacity and capability of an element to return to a pre-disturbance state after the influence of a disruption. Following the various definitions presented in Table 1.2, it is clear that regardless of context, the concept of resilience relates to achieving stability within the functioning of an element or system. The multifaceted concept is thereby composed of a complex set of interactions and behaviours. This allows the impacted element to recover and restore function. When the notion of resilience is then applied to communities and the wider context of organisations, this broad definition does not change dramatically. Resilience is related to both individual and organisational responses to turbulence and discontinuities. This involves both the ability to withstand systematic discontinuities as well as the capability to adapt to new risk environments (Starr et al. 2003).

Communities and organisations have both been conceptualised as complex systems by a variety of authors (Dooley 1997, Comfort et al. 2001, Crichton et al. 2009). A complex system is composed of interconnected elements forming a network of linkages that often interact non-linearly. This interaction gives rise to emergent behaviour as the system's interactions feedback within the network. This feedback creates a loop and reinforces the cause and effect relationship between system elements. In order to sustain the internal complexity of the system, constant energy and interaction between the system and the external environment is then required. This creates a dissipative system defined through the constant exchange of energy and matter.

Due to this behaviour, complex systems are evolutionary, reacting to local information and thereby capable of self-organisation (Andriani 2003). As identified by Comfort et al. (2001) when an environment's complexity increases, possibly through high impact or disruptive events, a system's performance decreases. The decrease occurs because the system is unable to process the amount and range of information required to adequately establish the coordination required across the components of the response system. This is a result of the system requiring a significant increase in information exchange, communication and coordination in order to integrate the multiple levels of system operation and decisions caused by the increase in environmental and system complexity. As a result, in order to establish a strategy for reducing risk in uncertain environments, Comfort et al. (2001) suggests that a system should create a balance between anticipation or preparedness and resilience. A consequence of this approach is that the system may evolve and learn. New structures, patterns and properties may then spontaneously emerge without being externally imposed by the system. This notion draws on many of the fundamental aspects of resilience as outlined by Holling (1973, 1996) within the context of ecological systems.

Within an ever-changing environment capable of significant turbulence, an organisational system is required to change and adapt in response to environmental fluctuations in order to sustain function and retain advantage. Without this change in the face of adversity, systems will follow a primarily recovery-based approach which may introduce maladaptive cycles of development. Instead, a resilience approach in the face of perturbation may enable an organisation to adapt to new risk environments and circumstances, providing a platform to effectively manage environmental variability and uncertainty (McDonald 2006). This adaption relates to the response of the organisation to threats and disruptive events and the ability to restore function. Through this, developing resilience within organisational elements provides a dynamic adjustment process recognising the complexity and non-linear causal relationships within the response of an organisation. Resilience thereby relates to the ability to recognise and adapt to threats and discontinuities, as well as developing the capability to rebound and effectively restore function following the impact of an event (Sutcliffe and Vogus 2003). As a result, organisational resilience is a complex concept that relates to the functioning of an organisation prior to, during and after the impact of an event.

1.5.1 VULNERABILITY

The term 'vulnerability' carries with it several connotations often connected to the susceptibility of a system to harm (Adger 2006). This view is reflected by McCarthy (2001) within the context of climate change. Here, vulnerability is defined as the degree to which a system is susceptible to the effects of climate change. Within a broader context, the concept of vulnerability is most often defined as an element being subject to a range of effects. These include exposure to perturbations, external stresses, sensitivity to perturbation (the degree to which the system is affected or altered due to perturbation) and the system's capacity of response (Gallopin 2006). Subsequently, the concept of vulnerability is inherently complex, impinging on several variables, and can therefore be thought of as a dynamic entity (Dalziell and McManus 2004).

Gallopin (2006) identifies the conceptual linkages between vulnerability, resilience and adaptive capacity. Within this representation, resilience is considered a subset or component of a system's capacity of response. A system's capacity of response relates to the ability of a system to adjust to a disturbance, moderate the effects, take advantage of any available opportunities and cope with the consequences of any system transformations (Gallopin 2006).

Within Gallopin's (2006) model, it is clear that vulnerability is the overreaching concept and that resilience and adaptive capacity are considered a conceptual subset. Through this, Gallopin (2006) refers to vulnerability as the capacity to preserve the structure of a system, while resilience refers to the capacity to recover from disturbances. The same relationship between vulnerability and resilience is reflected by Turner et al. (2003) within the development of vulnerability analysis models related to the concept of sustainability. Here, vulnerability is defined as the degree to which a system is likely to experience harm due to exposure to a threat or perturbation. As such, resilience is identified as a subset element of vulnerability (Figure 1.1).

1.5.2 ADAPTIVE CAPACITY

As shown by the examples of socio-ecological systems in a study by Carpenter et al. (2001) of resilience, the adaptive capacity of a system is related to the mechanisms for the creation of novelty and learning. Adaptive capacity is described by Gunderson

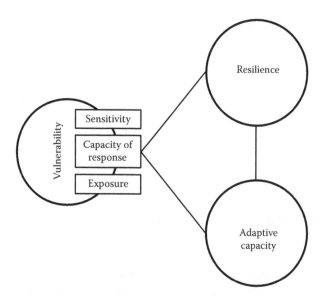

FIGURE 1.1 Concept of vulnerability. (Adapted from Gallopin, G.C., 2006. *Global Environmental Change*, 16 (3), 293–303.)

(2000) in regards to ecological resilience as a system's robustness to alterations and changes in resilience. Within Gallopin's (2006) model of the components of vulnerability, a system adaptive capacity is linked to a system capacity of response and defined as the ability of a system to evolve in order to accommodate environmental threats or changes and the ability to expand the range of variability.

Adaptive capacity reflects the ability of a system to respond to changes in its external environment and to recover from damage to internal structures within the system that affect the ability to achieve its purpose (Dalziell and McManus 2004). Predominately, literature has referred to or emphasised resilience as a means to recover from disturbances; however, the concept of adaptive capacity may also lead to establishing new system equilibriums or stability domains, allowing a system to adapt to new environments (Fiksel 2006). Here, resilience is established from a system of adaptive capacities and can be regarded as the process of linking resources to outcomes (Norris et al. 2008). As such, the adaptive capacity of a system can be regarded as the mechanism for resilience.

Carpenter et al. (2001) identify that the adaptive capacity of a system also reflects the learning aspect of system behaviour in response to a disruption. Within organisations, adaptive capacity refers to the ability to cope with unknown future circumstances (Staber and Sydow 2002), reflecting the definition set by Carpenter et al. (2001). As a result, organisations that focus on adaptive capacity will not experience environments passively. Instead, these organisations will continuously develop and apply new knowledge in relation to their operating environment. Rather than identifying the existing demands and then exploiting the available resources, adaptive organisations will reconfigure quickly in changing environments (Staber and Sydow

2002). Through this, the adaptive capacity of an organisation aids better preparedness for turbulent or uncertain environments.

1.5.3 RESILIENCE

Resilience is a function of both the vulnerability of a system and its adaptive capacity (Dalziell and McManus 2004). Resilience relates to the response of an element or system to impacts or turbulent conditions. In relation to the performance of an organisation, this adaption relates to the response of the organisation to threats and disruptive events as well as the ability to restore function. Implicit within this definition are four critical conditions: (1) exposure to a significant threat or event; (2) achievement of critical success factors within response activities; (3) capacity to alter function and operations and (4) ability to develop from experiences. Through this, developing resilience within organisational elements provides a dynamic adjustment process recognising the complexity and non-linear causal relationships within the response of an organisation.

This is reflected by Fiksel (2003) who identifies four major system characteristics that contribute to resilience. These are

- Diversity – the existence of multiple forms and behaviours
- Efficiency – performance with modest resource consumption
- Adaptability – flexibility to change in response to new pressures
- Cohesion – existence of unifying relationships and linkages between system variables and elements

To illustrate these characteristics, Fiksel (2003) presents simplified graphical representation of thermodynamic systems to characterise the different types of resilience. Each system has a stable state representing the lowest potential energy at which the system maintains order and function. When the system is subjected to a threat or perturbation, this state will shift along the trajectory of the adjacent states (Fiksel 2003). The examples of system behaviour are shown in Figure 1.2.

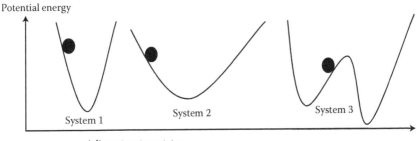

FIGURE 1.2   System trajectory. (Adapted from Fiksel, J., 2003. *Environmental Science and Technology*, 37 (23), 5330–5339.)

System 1 highlights an engineered system through which the system operates within a narrow band of possible states. Although the system is designed to be resistant to small disturbances from its equilibrium state, the system is unable to cope with larger scale or high impact events. As such the system may be regarded as resistant, but not as resilient. System 2 offers a greater resiliency to disturbances, as the system is able to retain fundamental function across a broad range of possible states and then gradually return to equilibrium. As a result system 2, typical of social and ecological systems, can be characterised as a resilient system. Although system 2 does classify as a resilient system, the characteristics of system 3 offer much greater resilience in the face of significant disturbance. Through the system having multiple equilibrium states, the system is able to shift to a different state under certain conditions. This means that the system is able to tolerate larger perturbations. However, the shift to a different equilibrium point represents a fundamental change in the systems structure and function (Fiksel 2003).

As highlighted by the various definitions of resilience presented in Table 1.2, dependent on the specific context, various perspectives have been applied to resilience. Whether focused on ecological, individual, organisational or infrastructural levels, the resilience narrative has developed a diverse literature base within addressing the response, recovery and adaption of elements and systems to threats and perturbations. However, the varying definitions and perspectives within this emerging literature base highlights the contrasting nature of the term.

Shaw and Maythorne (2013) define this contrast as either a focus on 'recovery' or 'transformation' within the resilience discourse. Through a 'recovery' driven focus emphasis is placed on resistance to external events and an efficient return to pre-disturbance operations as soon as possible. While the 'transformation' driven focus acknowledges that the impact of an event or disruption may be greater than the existing structures of a system are capable of withstanding. As a result, the impact of an event is not absorbed and the system is required to adapt in response to the impact or threat (Shaw and Theobald 2011). This perspective echoes the ecological foundations of the concept as outlined by Holling (1973).

Although several authors, such as Adger (2000), Carpenter et al. (2001), Walker et al. (2002), Folke et al. (2002), Starr et al. (2003), Sutcliffe and Vogus (2003) and Ponomarov and Holcomb (2009), recognise the ecological perspective of resilience, the relationship with organisational resilience is yet to be clearly defined. As highlighted by Raco and Street (2012), dependent on the context, varying perspectives have established their own selective narrative around the concept of resilience. Additionally, as highlighted by Manyena (2006), resilience-related concepts are also steeped in linguistic ambiguity, adding further complexity to the resilience narrative.

Resilience, as a concept, has a strong relationship with that of the notion of stability; this is reflected in Holling's (1973) original work with ecosystem stability. In addition, as shown by Gallopin (2006), resilience may also be conceptualised as a subset of vulnerability and a system's capacity of response. Although the conceptual linkages between this concept and resilience are recognised, there is still little empirical-based evidence to support the proposed relationship, particularly in relation to organisational level and systems resilience. However, while resilience and vulnerability may be viewed as factors of each other or separate entities, there is a need to adopt resilience thinking that extends beyond vulner-ability reduction and management (Manyena 2006). As a result, a more 'radical' approach has emerged emphasising a narrative of adaptation and transformation (Shaw and Maythorne 2013).

In relation to the increasing discussion around resilience, the concept has become a byword for the security response of agencies (Walker and Cooper 2011) and a metaphor within the policy-making process (Coaffee 2013). This is due to the politi-cal prioritisation of the safety and security within communities against potential threats and hazards; this focus has highlighted a greater requirement for foresight and preparedness in response to disruptive events (Coaffee 2013). Preparedness is achieved through the development of resilience within a system or community; this enhances the system's capacity to cope with disruptive events (Walker and Cooper 2011). Within this perspective, resilience is proactive rather than reactive (Coaffee 2013). However, as discussed by MacKinnon and Derickson (2013), the contested perspectives within the resilience narrative create a paradox of change; recognising the potential impact of turbulence and crises, yet accepting these events passively and placing the onus of response and adaption on the impacted systems or communities. To address this, MacKinnon and Derickson (2013) emphasise the importance of resourcefulness and the distribution of resources within and between communities.

As a result, it is important to recognise the context in which the concept of resil-ience is applied. While the term may be used to express the ability of a system or element to restore efficacy, resilience may also be used as a means to conceptualise adaption and performance within dynamic and complex environments. Therefore, it is important to identify the bounds within any narrative related to resilience and outline the foundations and positioning of any definition applied to it.

1.6   CONCLUSION AND FUTURE RESEARCH DIRECTIONS

From the evidence of this study there appears to be a strong focus around build-ing theories and definitions of resilience within the literature base. However, the literature is lacking in empirically proving the theories and exploring the underly-ing dynamics of resilience. As a result, there is limited critical discussion on how systems, such as organisations, can achieve greater degrees of resilience. For the theory to be of value in the real world, further exploration and research needs to be conducted, particularly focused on applying empirical methods such as case study and survey methodologies which can significantly add to and validate theoretical constructs. Although a number of case studies are found within the literature, only a relatively small number focus on organisational resilience. Empirical research that uses case-based methods focusing on the organisation are currently limited and further study is required to improve current understanding.

Some areas of resilience have received significant academic attention and empiri-cal study, such as ecological systems (Holling 1973, Gunderson 2000, Carpenter et al. 2001) and to a lesser extent socio-ecological systems (Walker et al. 2002, 2004). Accordingly, areas such as organisational level resilience and particularly the extended enterprise and supply chains need a greater focus. There is also need to conduct good quality empirical-based research to fully develop the area and properly recognise the potential of developing resilient characteristics within organisations and supply chains. The supply chain is a context that is currently well served, particularly in the area of risk and its avoidance.

Considering the various definitions of 'resilience' from the literature and as depicted in Table 1.2, it is essential to understand whether resilience is a measure, a feature, a philosophy or a capability? Again, is being resilient a tangible capability or an intangible capability? Although these may be difficult questions to ground, the authors believe that this will largely depend upon the context: individual, organisational, supply chain, community and ecological. In some cases, there may be a need for the research to transcend all these contexts. Researchers must untangle the complexities involved.

REFERENCES

Adger, W.N., 2000. Social and ecological resilience: Are they related? *Progress in Human Geography*, 24 (3), 347.

Adger, W.N., 2006. Vulnerability. *Global Environmental Change*, 16 (3), 268–281.

Ainuddin, S. and Routray, J.K., 2012. Community resilience framework for an earthquake prone area in Baluchistan. *International Journal of Disaster Risk Reduction*, 2, 25–36.

Akahane, K., Yonai, S., Fukuda, S., Miyahara, N., Yasuda, H., Iwaoka, K. and Akashi, M., 2012. The Fukushima Nuclear Power Plant accident and exposures in the environment. *The Environmentalist*, 32 (2), 136–143.

Aleksić, A., Stefanović, M., Arsovski, S. and Tadić, D., 2013. An assessment of organizational resilience potential in SMEs of the process industry, a fuzzy approach. *Journal of Loss Prevention in the Process Industries*, 26 (6), 1238–1245.

Alexander, D., 2003. Terrorism, disasters, and security. *Prehospital and Disaster Medicine*, 18 (3), 165–169.

Allenby, B. and Fink, J., 2005. Toward inherently secure and resilient societies. *Science*, 309, 1034–1036.

Andriani, P., 2003. Evolutionary dynamics of industrial clusters. In: E. Mitleton-Kelly (ed.), *Complex Systems and Evolutionary Perspectives on Organisations: The Application of Complexity Theory to Organisations*. Oxford: Elsevier Science, pp. 127–145.

Barnett, C.K. and Pratt, M.G., 2000. From threat-rigidity to flexibility: Toward a learning model of autogenic crisis in organizations. *Journal of Organizational Change Management*, 13 (1), 74–88. BIS (2010). [Online] URL: http://stats.bis.gov.uk/ed/sme/. [Accessed 2/11/2010].

Bodin, P. and Wiman, B., 2004. Resilience and other stability concepts in ecology: Notes on their origin, validity, and usefulness. *ESS Bulletin*, 2 (2), 33–43.

Bonanno, G.A., 2004. Loss, trauma, and human resilience. *American Psychologist*, 59 (1), 20–28.

Bonanno, G.A. et al., 2006. Psychological resilience after disaster. *Psychological Science*, 17 (3), 181.

Braes, B. and Brooks, D., 2011. Organisational resilience: Understanding and identifying the essential concepts. *Safety and Security Engineering IV*, 117, 117.

Brand, F., 2009. Critical natural capital revisited: Ecological resilience and sustainable development. *Ecological Economics*, 68 (3), 605–612.

Bruneau, M. et al., 2003. A framework to quantitatively assess and enhance the seismic resilience of communities. *Earthquake Spectra*, 19 (4), 733–752.

Burnard, K. and Bhamra, R., 2011. Organisational resilience: Development of a conceptual framework for organisational responses. *International Journal of Production Research*, 49 (18), 5581–5599.

Callister, W.D., 2003. Mechanical properties of metals. *Materials Science and Engineering: An Introduction*, 6th edition. New York: John Wiley and Sons, Inc., pp. 129–130.

Carpenter, S. et al., 2001. From metaphor to measurement: Resilience of what to what? *Ecosystems*, 4 (8), 765–781.

Cerullo, V. and Cerullo, M.J., 2004. Business continuity planning: A comprehensive approach. *Information Systems Management*, 21 (3), 70–78.

Chen, C.H. et al., 2007. Long-term psychological outcome of 1999 Taiwan earthquake survivors: A survey of a high-risk sample with property damage. *Comprehensive Psychiatry*, 48 (3), 269–275.

Christopher, M. and Peck, H., 2004. Building the resilient supply chain. *International Journal of Logistics Management*, 15 (2), 1–14.

Coaffee, J., 2013. Rescaling and responsibilising the politics of urban resilience: From national security to local place-making. *Politics*, 33 (4), 240–252.

Comfort, L.K. et al., 2001. Complex systems in crisis: Anticipation and resilience in dynamic environments. *Journal of Contingencies and Crisis Management*, 9 (3), 144–158.

Coutu, D.L., 2002. How resilience works. *Harvard Business Review*, 80 (5), 46–56.

Crichton, M.T. et al., 2009. Enhancing organizational resilience through emergency planning: Learnings from cross-sectoral lessons. *Journal of Contingencies and Crisis Management*, 17 (1), 24–37.

Cumming, G.S. et al., 2005. An exploratory framework for the empirical measurement of resilience. *Ecosystems*, 8 (8), 975–987.

Dalziell, E.P. and McManus, S.T., 2004. Resilience, vulnerability, and adaptive capacity: Implications for system performance. *International Forum for Engineering Decision Making (IFED)*. Christchurch: University of Canterbury.

Derissen, S., Quaas, M.F. and Baumgärtner, S., 2011. The relationship between resilience and sustainability of ecological-economic systems. *Ecological Economics*, 70 (6), 1121–1128.

Dooley, K.J., 1997. A complex adaptive systems model of organization change. *Nonlinear Dynamics, Psychology, and Life Sciences*, 1 (1), 69–97.

Easterby-Smith, M., Thorpe, R. and Lowe, A., 2002. *Management Research: An Introduction*, second edition. London: Sage Publications Ltd.

Elmqvist, T. et al. 2003. Response diversity, ecosystem change, and resilience. *Frontiers in Ecology and the Environment*, 1 (9), 488–494.

Engle, N.L., 2011. Adaptive capacity and its assessment. *Global Environmental Change*, 21 (2), 647–656.

Falasca, M., Zobel, C.W. and Cook, D., 2008. A decision support framework to assess supply chain resilience. *Paper Presented at International ISCRAM Conference*, May, Washington, DC.

Fiksel, J., 2003. Designing resilient, sustainable systems. *Environmental Science and Technology*, 37 (23), 5330–5339.

Fiksel, J., 2006. Sustainability and resilience: Toward a systems approach. *Sustainability: Science Practice and Policy*, 2 (2), 14–21.

Folke, C., Carpenter, S., Elmqvist, T., Gunderson, L., Holling, C.S. and Walker, B., 2002. Resilience and sustainable development: Building adaptive capacity in a world of transformations. *AMBIO: A Journal of the Human Environment*, 31 (5), 437–440.

Folke, C., Carpenter, S.R., Walker, B., Scheffer, M., Chapin, T. and Rockström, J., 2010. Resilience thinking: Integrating resilience, adaptability and transformability. *Ecology and Society*, 15 (4), 20.

Gallopin, G.C., 2006. Linkages between vulnerability, resilience, and adaptive capacity. *Global Environmental Change*, 16 (3), 293–303.

Garmezy, N. and Masten, A.S., 1986. Stress, competence, and resilience: Common frontiers for therapist and psychopathologist. *Behavior Therapy*, 17 (5), 500–521.

Gibson, C.A. and Tarrant, M., 2010. A 'Conceptual Models' approach to organisational resilience. *Australian Journal of Emergency Management*, 25 (2), 6.

Goodman, P.S., Ramanujam, R., Carroll, J.S., Edmondson, A.C., Hofmann, D.A. and Sutcliffe, K.M., 2011. Organizational errors: Directions for future research. *Research in Organizational Behavior*, 31, 151–176.

Gunderson, L.H., 2000. Ecological resilience-in theory and application. *Annual Review of Ecology and Systematics*, 31 (1), 425–439.

Hamel, G. and Valikangas, L., 2003. The quest for resilience. *Harvard Business Review*, 81 (9), 52–65.

Harlow, W.F., Brantley, B.C. and Harlow, R.M., 2011. BP initial image repair strategies after the Deepwater Horizon spill. *Public Relations Review*, 37 (1), 80–83.

Harwood, I., Humby, S. and Harwood, A., 2011. On the resilience of corporate social responsibility. *European Management Journal*, 29 (4), 283–290.

Hawes, C. and Reed, C., 2006. Theoretical steps towards modelling resilience in complex systems. *Computational Science and Its Applications-ICCSA 2006*, 3980, 644–653.

Henry, D. and Ramirez-Marquez, J.E., 2012. Generic metrics and quantitative approaches for system resilience as a function of time. *Reliability Engineering and System Safety*, 99, 114–122.

Hind, P., Frost, M. and Rowley, S., 1996. The resilience audit and the psychological contract. *Journal of Managerial Psychology*, 11 (7), 18–29.

Hodbod, J. and Adger, W.N., 2014. Integrating social-ecological dynamics and resilience into energy systems research. *Energy Research and Social Science*, 1, 226–231.

Holling, C.S., 1973. Resilience and stability of ecological systems. *Annual Review of Ecology and Systematics*, 4 (1), 1–23.

Holling, C.S., 1996. Engineering resilience versus ecological resilience. In: P.C. Schulze (ed.), *Engineering within Ecological Constraints*. Washington, DC: National Academy of Engineering, pp. 31–43.

Holling, C.S., 2001. Understanding the complexity of economic, ecological, and social systems. *Ecosystems*, 4 (5), 390–405.

Hollnagel, E., Woods, D.D. and Leveson, N., 2006. *Resilience Engineering: Concepts and Precepts*. Aldershot: Ashgate.

Horne, J.F. and Orr, J.E., 1998. Assessing behaviors that create resilient organizations. *Employment Relations Today*, 24, 29–40.

Hughes, T.P. et al., 2005. New paradigms for supporting the resilience of marine ecosystems. *Trends in Ecology and Evolution*, 20 (7), 380–386.

Johnson, N., Elliott, D. and Drake, P., 2013. Exploring the role of social capital in facilitating supply chain resilience. *Supply Chain Management: An International Journal*, 18 (3), 324–336.

Juttner, U., 2005. Supply chain risk management. *International Journal of Logistics Management*, 16 (1), 120–141.

Jüttner, U. and Maklan, S., 2011. Supply chain resilience in the global financial crisis: An empirical study. *Supply Chain Management: An International Journal*, 16 (4), 246–259.

Kachali, H., Stevenson, J.R., Whitman, Z., Seville, E., Vargo, J. and Wilson, T., 2012. Organisational resilience and recovery for Canterbury organisations after the 4 September 2010 earthquake. *Australasian Journal of Disaster and Trauma Studies*, 1, 11–19.

Klein, R.J.T., Nicholls, R.J. and Thomalla, F., 2003. Resilience to natural hazards: How useful is this concept? *Environmental Hazards*, 5 (1–2), 35–45.

Lengnick-Hall, C.A., Beck, T.E. and Lengnick-Hall, M.L., 2011. Developing a capacity for organizational resilience through strategic human resource management. *Human Resource Management Review*, 21 (3), 243–255.

Linnenluecke, M. and Griffiths, A., 2010. Beyond adaptation: Resilience for business in light of climate change and weather extremes. *Business and Society*, 49 (3), 477–511.

Luthans, F., Vogelgesang, G.R. and Lester, P.B., 2006. Developing the psychological capital of resiliency. *Human Resource Development Review*, 5 (1), 25.

Luthar, S.S., Cicchetti, D. and Becker, B., 2000. The construct of resilience: A critical evaluation and guidelines for future work. *Child Development*, 71 (3), 543–562.

MacKinnon, D. and Derickson, K.D., 2013, From resilience to resourcefulness: A critique of resilience policy and activism. *Progress in Human Geography*, 37 (2), 253–270.

Mallak, L., 1998. Putting organizational resilience to work. *Industrial Management*, 40, 8–13.

Manyena, S.B., 2006. The concept of resilience revisited. *Disasters*, 30 (4), 434–450.

Masten, A.S., Best, K.M. and Garmezy, N., 1990. Resilience and development: Contributions from the study of children who overcome adversity. *Development and Psychopathology*, 2 (4), 425–444.

Masten, A.S. and Coatsworth, J.D., 1998. The development of competence in favorable and unfavorable environments. *American Psychologist*, 53 (2), 205–220.

Masten, A.S. and Obradovic, J., 2007. Disaster preparation and recovery: Lessons from research on resilience in human development. *Ecology and Society*, 13 (1), 9. [online] URL: http:// www.ecologyandsociety.org/vol13/iss1/art9/ [Accessed 8/10/2010].

McCarthy, J.J., 2001. Climate change 2001: Impacts, Adaptation, and Vulnerability. In: *Contribution of Working Group II to the Third Assessment Report of the Intergovernmental Panel on Climate Change*. Cambridge: Cambridge University Press, pp. 21–22.

McDonald, N., 2006. Organisational resilience and industrial risk. In: E. Hollnagel, D.D. Woods and N. Leveson (eds.), *Resilience Engineering: Concepts and Precepts*. Hampshire: Ashgate, pp. 155–179.

McManus, S. et al., 2007. *Resilience Management: A Framework for Assessing and Improving the Resilience of Organisations*. Christchurch, New Zealand: Resilient Organisations Research Programme.

Miller, D. and Whitney, J.O., 1999. Beyond strategy: Configuration as a pillar of competitive advantage. *Business Horizons*, 42 (3), 5–17.

Mitchell, R.J. et al., 2000. Ecosystem stability and resilience: A review of their relevance for the conservation management of lowland heaths. *Perspectives in Plant Ecology, Evolution and Systematics*, 3 (2), 142–160.

Muralidharan, S., Dillistone, K. and Shin, J.H., 2011. The Gulf Coast oil spill: Extending the theory of image restoration discourse to the realm of social media and beyond petroleum. *Public Relations Review*, 37 (3), 226–232.

Nelson, D.R., Adger, W.N. and Brown, K., 2007. Adaptation to environmental change: Contributions of a resilience framework. *Annual Review of Environment and Resources*, 32 (1), 395.

Norris, F.H. et al., 2008. Community resilience as a metaphor, theory, set of capacities, and strategy for disaster readiness. *American Journal of Community Psychology*, 41 (1), 127–150.

Norrman, A. and Jansson, U., 2004. Ericsson's proactive supply chain risk management approach after a serious sub-supplier accident. *International Journal of Physical Distribution and Logistics Management*, 34 (5), 434–456.

Nystrom, M., Folke, C. and Moberg, F., 2000. Coral reef disturbance and resilience in a human-dominated environment. *Trends in Ecology and Evolution*, 15 (10), 413–417.

Ong, A.D. et al., 2006. Psychological resilience, positive emotions, and successful adaptation to stress in later life. *Journal of Personality and Social Psychology*, 91 (4), 730.

Pal, R., Torstensson, H. and Mattila, H., 2014. Antecedents of organizational resilience in economic crises—An empirical study of Swedish textile and clothing SMEs. *International Journal of Production Economics*, 147, 410–428.

Papadakis, I.S., 2006. Financial performance of supply chains after disruptions: An event study. *Supply Chain Management: An International Journal*, 11 (1), 25–33.

Paton, D. and Johnston, D., 2001. Disasters and communities: Vulnerability, resilience and preparedness. *Disaster Prevention and Management: An International Journal*, 10 (4), 270–277.

Paton, D., Smith, L. and Violanti, J., 2000. Disaster response: Risk, vulnerability and resilience. *Disaster Prevention and Management*, 9 (3), 173–180.

Pereira, C.R., Christopher, M. and Lago Da Silva, A., 2014. Achieving supply chain resilience: The role of procurement. *Supply Chain Management: An International Journal*, 19 (5/6), 626–642.

Petchey, O.L. and Gaston, K.J., 2009. Effects on ecosystem resilience of biodiversity, extinctions, and the structure of regional species pools. *Theoretical Ecology*, 2 (3), 177–187.

Peterson, G., Allen, C.R. and Holling, C.S., 1998. Ecological resilience, biodiversity, and scale. *Ecosystems*, 1 (1), 6–18.

Ponomarov, S.Y. and Holcomb, M.C., 2009. Understanding the concept of supply chain resilience. *International Journal of Logistics Management*, 20 (1), 124–143.

Poortinga, W., 2012. Community resilience and health: The role of bonding, bridging, and linking aspects of social capital. *Health and Place*, 18 (2), 286–295.

Powley, E.H., 2009. Reclaiming resilience and safety: Resilience activation in the critical period of crisis. *Human Relations*, 62 (9), 1289.

Raco, M. and Street, E., 2012. Resilience planning, economic change and the politics of post-recession development in London and Hong Kong, *Urban Studies,* 49 (5), 1065–1087.

Reich, J.W., 2006. Three psychological principles of resilience in natural disasters. *Disaster Prevention and Management*, 15 (5), 793–798.

Rice, J.B. and Sheffi, Y., 2005. A supply chain view of the resilient enterprise. *MIT Sloan Management Review*, 47 (1), 41.

Riolli, L. and Savicki, V., 2003. Information system organizational resilience. *Omega*, 31 (3), 227–233.

Rudolph, J.W. and Repenning, N.P., 2002. Disaster dynamics: Understanding the role of quantity in organizational collapse. *Administrative Science Quarterly*, 47 (1), 1–30.

Ruiz-Ballesteros, E., 2011. Social-ecological resilience and community-based tourism: An approach from Agua Blanca, Ecuador. *Tourism Management*, 32 (3), 655–666.

Rutter, M., 2006. Implications of resilience concepts for scientific understanding. *Annals of the New York Academy of Sciences*, 1094 (1), 1–12.

Scholten, K., Sharkey Scott, P. and Fynes, B., 2014. Mitigation processes–antecedents for building supply chain resilience. *Supply Chain Management: An International Journal*, 19 (2), 211–228.

Seville, E. et al., 2006. *Building Organisational Resilience: A Summary of Key Research Findings*. Christchurch, New Zealand: Resilient Organisations Research Programme.

Shaw, K. and Maythorne, L., 2013. Managing for local resilience: Towards a strategic approach. *Public Policy and Administration,* 28 (1), 43–65.

Shaw, K. and Theobald, K., 2011. Resilient local government and climate change interventions in the UK. *Local Environment,* 16 (1), 1–15.

Sheffi, Y., 2005. Building a resilient supply chain. *Harvard Business Review Supply Chain Strategy*, 1 (5), 1–11.

Sheffi, Y., 2007. *The Resilient Enterprise: Overcoming Vulnerability for Competitive Advantage*. Cambridge, MA: MIT Press.

Smith, D. and Fischbacher, M., 2009. The changing nature of risk and risk management: The challenge of borders, uncertainty and resilience. *Risk Management*, 11 (1), 1–12.

Srivastava, S.K., 2007. Green supply-chain management: A state-of the-art literature review. *International Journal of Management Reviews*, 9 (1), 53–80.

Staber, U. and Sydow, J., 2002. Organizational adaptive capacity: A structuration perspective. *Journal of Management Inquiry*, 11 (4), 408.

Starr, R., Newfrock, J. and Delurey, M., 2003. Enterprise resilience: Managing risk in the networked economy. *Strategy and Business*, 30, 70–79.

Steen, R. and Aven, T., 2011. A risk perspective suitable for resilience engineering. *Safety Science*, 49 (2), 292–297.

Stevenson, M. and Busby, J., 2015. An exploratory analysis of counterfeiting strategies: Towards counterfeit-resilient supply chains. *International Journal of Operations and Production Management*, 35 (1), 110–144.

Sutcliffe, K.M. and Vogus, T.J., 2003. Organizing for resilience. In: K.S. Cameron, J.E. Dutton and R.E. Quinn (eds.), *Positive Organizational Scholarship: Foundations of a New Discipline*. San Francisco, CA: Berrett-Koehler, pp. 94–110.

Tang, C.S., 2006. Robust strategies for mitigating supply chain disruptions. *International Journal of Logistics Research and Applications*, 9 (1), 33–45.

Thomson, D.A. and Lehner, C.E., 1976. Resilience of a rocky intertidal fish community in a physically unstable environment. *Journal of Experimental Marine Biology and Ecology*, 22 (1), 1–29.

Tilman, D. and Downing, J.A., 1994. Biodiversity and stability in grasslands. *Nature*, 367, 363–365.

Turner, B.L. et al., 2003. A framework for vulnerability analysis in sustainability science. *Proceedings of the National Academy of Sciences of the United States of America*, 100 (14), 8074.

Tveiten, C.K., Albrechtsen, E., Wærø, I. and Wahl, A.M., 2012. Building resilience into emergency management. *Safety Science*, 50 (10), 1960–1966.

Vis, M. et al., 2003. Resilience strategies for flood risk management in the Netherlands. *International Journal of River Basin Management*, 1 (1), 33–40.

Vogus, T.J. and Sutcliffe, K.M., 2007. Organizational resilience: Towards a theory and research agenda. In: *IEEE International Conference on Systems, Man and Cybernetics*, ISIC, 7–10 October, Montreal, 3418.

Walker, B. et al., 2002. Resilience management in social–ecological systems: A working hypothesis for a participatory approach. *Conservation Ecology*, 6 (1), 14.

Walker, B. et al., 2004. Resilience, adaptability and transformability in social–ecological systems. *Ecology and Society*, 9 (2), 5.

Walker, J. and Cooper, M., 2011. Genealogies of resilience: From systems ecology to the political economy of crisis adaptation. *Security Dialogue*, 42 (2), 143–160.

Waters, D., 2007. *Supply Chain Risk Management: Vulnerability and Resilience in Logistics*. London: Kogan Page.

Werner, J., 2009. Risk and risk aversion when states of nature matter. *Economic Theory*, 41 (2), 231–246.

Wilkinson, C., 2012. Social-ecological resilience: Insights and issues for planning theory. *Planning Theory*, 11 (2), 148–169.

Youssef, C.M. and Luthans, F., 2007. Positive organizational behavior in the workplace: The impact of hope, optimism, and resilience. *Journal of Management*, 33 (5), 774.

2 The Ideal of Resilient Systems and Questions of Continuity

Uta Hassler and Niklaus Kohler

CONTENTS

2.1 RESILIENCE AND THE BUILT ENVIRONMENT

The notion of 'built environment' has been in common use since the mid-1970s. It refers to the man-made landscapes that provide the setting for human activity, ranging from the large-scale urban entities to personal dwelling places. The term responds to the need of a multitude of actors and professions to find a common framework for communication and elaboration.[*] The origin of the notion clearly resides in anthropological and behavioural studies about the influence of form and space on the individual and on social behaviour (Rapoport, 1976). In more recent research, the built environment is understood as the result of a process of social construction (Lawrence and Low, 1990, p. 455). Developments in system ecology and environmental economics have led to definitions of the built environment formulated in relation to the 'non-built' environment, that is, the ecosphere, constituting a complex, dynamic, self-producing system.

The demarcation between the built environment and the ecosystem does not exist as such; it is historically constructed and therefore changing. Fischer-Kowalski and Weisz (1999) have synthesised a transdisciplinary framework that facilitates a unified theory. It incorporates two key concepts: the socio-economic metabolism and the colonisation of natural processes. Metabolism refers to the balanced flows

---

[*] A definition of the built environment has been presented more in detail by Moffatt and Kohler (2008).

of energy and materials between the human and natural sub-systems of the material realm. Colonisation describes the appropriation by culture of elements of the material realm in order to assure the reproduction (and possibly the expansion) of a society. The built environment (as an artefact) is in an overlapping zone between culture and nature, with causation occurring in both directions (Fischer-Kowalski and Weisz, 1999, p. 242).

Buildings, infrastructures and cultural landscapes* constitute the built environment. They can be considered as a set of natural, physical, economic, human, social and cultural capitals. In a sustainable paradigm the emphasis is on maintaining (sustaining) the value of all these capitals. However, natural and cultural capitals have a unique position because they cannot be substituted or replaced. Based on these limiting assumptions, the trade-offs between different strategies can be realised through the search of a Pareto frontier between the different capital stocks (Hoffenson et al., 2013).

In spite of considerable efforts to implement a sustainable management of the built environment, there is no general agreement on the concepts and tools to cope with its multiscalar character and the very different time scales. A number of concepts are used to qualify the (sustainable) development of the built environment and to inform decisions on the management of buildings and infrastructures. Continuity and duration refer to stability and different forms of equilibrium. Durability and robustness are based on the control of vulnerability. Loss of value can result from exposure to different types of risks. These risks can be characterised as fast disruptance or as slow, gradual change (i.e. obsolescence and economic decline). The concept of resilience relates in different ways to these concepts, particularly when it addresses the question of change and continuity over time. A key question is whether resilience is a useful additional concept alongside various concepts relating to 'sustainability'. If this is the case, then it must be possible to describe how resilience can be implemented. Is resilience a new meta-concept (the basis of a new heuristic)? Or is it only a useful metaphor (or a boundary object) when dealing with the built environment in a transdisciplinary perspective?

This chapter is structured as follows. It starts with an exploration into the possibility of establishing continuity in the form of alternative types of equilibria and the search for indicators for the survival characteristics of buildings and infrastructure. The questions of duration and vulnerability are then analysed as part of 'long foresight' strategies. Robustness is related to experiences from history. Fast- and slow-moving risks are analysed in relation to different planning and anticipation strategies. The double aspects of cultural capital (i.e. material and intangible) are discussed in relation to the different time effects. Finally, the built environment is presented as a 'depository of values' that is still subject to comparably slow changes. It is argued that natural and cultural capital can only be maintained if they are conservatively used and if adaptation occurs slowly. This highlights the central importance of resilience as a timing tool.

---

* The term 'cultural landscape' (*Kulturlandschaft*) in its largest definition goes back to Sauer (1925, p. 46): 'The cultural landscape is fashioned from a natural landscape by a cultural group. Culture is the agent, the natural area is the medium, the cultural landscape is the result'.

2.2 CONTINUITY, STABILITY AND STEADY-STATE EQUILIBRIA

The definition of sustainability is basically a set of goals that implicitly refers to the maintenance of equilibrium between the built and the natural environment, that is, between culture and nature. This equilibrium depends essentially on the kind of changes and on the adopted time scale. In his essay on the future of ecology Reichholf (2008) discusses how change is inevitable and how anthropogenic factors intervene. An equilibrium is only possible in the form of 'stable disequilibria' (*stabile Ungleichgewichte*). In socio-ecological systems, long cycles of constant care and cultivation allow a 'continuum of stable disequilibria' at best. This type of equilibrium is not necessarily stable over the long-term, it corresponds more to a lively river than to a quiet lake. Nature has developed 'copy processes' of species in which series, similarities and regularities can occur. Certain communities of species will only begin a process of inhabitation of a particular place after hundreds of years. On the other hand, breaches, instabilities and collapses also happen which can lead to the development of changed and new systems that are of different complexity and quality. Stewardship (i.e. constant care, intelligent usage and a view toward transmission to future generations) promotes thinking of a (longer) time duration.

When taxonomical orders have given way to historical orders,[*] the idea that a history of development shapes the world is essential to gain insight into the 'historicity of natural things'. Such a notion also relies upon the positivist idea of 'progress' (Rapp, 1992) to underpin it: the belief that technical systems can be controlled and that all social concerns can be optimised using knowledge and science. These two aspects form the basis of this figure of thought. The '*querelles des anciens et modernes*' (quarrel of the Ancients and the Moderns) and the (French) Enlightenment have promoted the ideal of improvement through new science. This approach to cultural history implicitly postulates a need to develop further the 'possibilities of modernity'. The certitude that the potential exists systematically to expand and improve knowledge also requires the development of the cultural and built systems of a society. The ambitions to dominate nature, to optimise the possibilities of human life and to develop oneself by means of scientific knowledge are both incentives and bold expectations of a better life.

Scientific curiosity often stimulates new classifications of knowledge. Cultural acts (breeding, husbandry and selection) change objects, things of nature and artefacts. The perception that a history of development is a 'success story' of evolution suggests in many ways the belief in the 'better new'. During the mid-twentieth century, experts in architecture and city planning propagated normative perceptions of a better future that resulted from relinquishing and replacing the existing built environment. At that time, the consequences of war were discussed as an opportunity for a better new beginning. However, from the perspective of the early twenty-first century, a breakdown of technical and cultural systems can be observed: the 'negative developments' of big investment decisions and the (long-term) consequences of human-made and natural catastrophes. All these observations again suggest a strong connection between time and the consequences of risks.

---

[*] For an overview of theoretical concepts of history, cf. Küttler et al. (1994).

Historical research on the general frameworks and possibilities of ancient civilisations demonstrates that the continuous transmission of knowledge across several generations is essential for cultural development. As the Egyptologists Jan and Aleida Assmann (Assmann, A., 1999; Assmann, J., 1999) argued, societies do need artefacts as an aid that allows them to 'remember' and to orientate themselves between generations. In his writings on 'cultural memory' Jan Assmann (1999, 2010) further developed Claude Lévi-Strauss's famous discussion on 'cold' and 'hot' societies. 'Peoples without history'* (Lévi-Strauss, 1966), that is, 'cold' societies stand in opposition to 'hot' societies. Such cold societies are equipped with a 'particular wisdom' to 'desperately resist any changes in their structure that could allow an intrusion of history'. On the other hand, 'hot societies' are 'avid for change' (Assmann, 1999, p. 68, 2010). Nevertheless, Jan Assmann (1999) proposes a reinterpretation of Lévi-Strauss's verbal images. In Assmann's view, the various cultural ways of how to deal with memory and history show a consciousness of continuity and regularity or even the 'singular, exceptional'. This should be separately discussed from breaches and changes (p. 70). Narratives of history serve as an 'engine to development' or become even the 'foundation of continuity' (p. 70). Cultures that are successful in the long run have both structures of purposeful 'iterative cultivation of traditions' (how to remember) and concepts of how to develop cultural skills further (how to learn). The relation of remembering and learning forms a type of dynamic equilibrium that corresponds probably to the definition of 'ecosystem resilience', of a process where change does not lead to the return to a single defined stable state, but can lead to different stable states.† Absolute stability is therefore neither possible nor desirable, but some indicators point to structures that are able to change and survive. These indicators also show that there exist traditional (implicit) institutional regimes that aim to prevent hazards: land reserves, water regimes, regimes of usage from the three-field crop rotation to the irrigation of pastures, redundant constructions of buildings and infrastructures, etc. In fact, the historic 'European city' as such shows a high rate of survival. Compared with other continents, few European cities have disappeared. Evidence from present efforts to cope with the coming climate change shows that the response at a city level is often much more consistent than corresponding national efforts.‡

---

* 'I have suggested elsewhere that the clumsy distinction between "peoples without history" and others could be fruitfully replaced by the distinction between what for convenience I called "cold" and "hot" societies: the former seeking through the institutions they gave themselves to annul the possible effects of historical factors on their equilibrium and continuity in a quasi-automatic fashion; the latter resolutely internalizing the historical process and making it the moving power of their development' (Lévi- Strauss, 1966, pp. 233–234).

† Crawford Stanley Holling gave the following initial definition: 'The first definition […] concentrates on stability near an equilibrium steady-state, where resistance to disturbance and speed of return to the equilibrium are used to measure the property' (Holling, 1996, p. 33). Based on this, the Resilience Alliance distinguishes between 'engineering resilience' (near a static equilibrium, but in the domain of a dynamic, steady-state, equilibrium with the efficiency of function as an objective) and 'ecosystem resilience' (remote from a static equilibrium with the existence of function in focus) (Holling, 1996, p. 33).

‡ See the International Council for Local Environmental Initiatives (ICLEI) network on Local Governments for Sustainability (2012). It proposes as an objective a resilient city: 'A resilient city is low risk to natural and man-made disasters. It reduces its vulnerability by building on its capacity to respond to climate change challenges, disasters and economic shocks' (http://www.iclei.org).

2.3   VULNERABILITY AND THE DURATION OF THE BUILT ENVIRONMENT

The built environment belongs to the kind of artefact that can survive over a long period of time. Until the end of the nineteenth century, the common ideal was durability and continued use. The deliberate shift to 'short-term products' was a consequence of industrial manufacturing techniques and led to short product cycles. Effectively short product cycles transfer the risk to future generations which poses relevant questions about resilient systems. After roughly a century of industrial economic activity in the construction industry, new vulnerabilities of built systems have gradually appeared (Hassler and Kohler, 2004). These are, among others

- End of the life span for nineteenth-century infrastructure (water, sewage systems, railways, etc.) has created difficulties to afford their renewal
- Shortened life expectations of building equipment (energy supply, building services and mechanical systems, etc.)
- Rising costs of building operation (due to higher standards, codes, demands, etc.)
- Pollutants that are dispersed in the built environment (partially through recycling)
- Increased risks from the adoption of uniform, generalised technical solutions, 'disasters by design' (Mileti, 1999)
- Lack of robust and redundant solutions and fault-tolerant systems (increasing the amount of irreparable constructions)
- Internationalised markets with regional externalisation of costs

It is becoming evident in many fields that the traditional mechanisms of optimisation have reached their limits (Rapp, 1994). The insight that possible 'best practices' in history cannot be reproduced is evident within the fields of building history and conservation science. All these phenomena show an increase in vulnerability that is often (but not always) accompanied by loss of resilience. The attempt to increase resilience principles in engineering and risk management raises the question of time. In the engineering definition of resilience, the control of time is fundamental: 'in order to be in control it is necessary to know what happened (the past) what happens (the present) and may happen (the future)' (Hollnagel et al., 2007). In this interpretation, resilience would be a superior guiding principle that encompasses the experience of long-term surviving systems but that also encompasses the limited ability to make predictions. Resilience would become a central, strategic 'timing tool'.[*]

---

[*] Cf. the notions of time and timing in Norbert Elias's view: 'Timing thus is based on people's capacity for connecting with each other two or more different sequences of continuous changes, one of which serves as a timing standard for the other (or others)' (Elias, 1992, p. 72). The social construction of time, therefore, goes back to a specific human ability to work on the experience of change, to react, to organise and confer meaning on the experience.

A new debate on resilience needs to examine how to differentiate the various risk dynamics but also must include a discussion on the quality of the losses.

The reduction of vulnerability through the possibility of adaptation of material structures (in particular in the form of real options) depends on the readiness for 'extended foresight'. Extended foresight embraces both material aspects as well as cultural value systems and intangible goods. Returning to the specific definitions of resilience, the long-term behaviour of the built environment depends on three types of resilience (Folke, 2006):

- Engineering resilience as material property of the fabric
- Ecosystem resilience as the buffer capacity allowing the persistence of the different component of the system
- 'Adaptive capacity' processes within a social–ecological system or 'adaptive creativity' in the case of culture

However, the concept of resilience changes according to the scale from an engineering definition on the level of buildings (stability) to an ecosystem-type definition when addressing the neighbourhood, urban and regional levels. The temporal uncertainty grows with the scale of the environment just as the capacity to react (the resilience) changes from material properties to the capacity to adapt and to learn and finally leads to long-term cultural changes.

2.4 'ROBUSTNESS': LESSONS FROM HISTORY

Before the advent of industrial principles of construction in the nineteenth century, it was known and accepted that constructions could be used for a long period of time if regularly maintained. The 'robustness' of these constructions is the result of their material properties and their skillful and regular maintenance. In history, 'robust' constructions were based on

- High material input, monolithic assembly configuration, extraordinary weight of single parts, use of 'everyday' materials, precise assembly solutions (e.g. the great stone works of antiquity, water management systems and bridges)
- Properties of reparability and exchangeability of (mostly identical or serialised) single parts of a well-proven construction (the tiled roof, constructions in wood)
- Neutrality of use (including planned redundancy)
- Oversized individual parts
- Self-evident systems (clarity of assembly, clarity of structure, etc.)
- Established systems of experts (craftsmen) spanning generations (i.e. the transmission of skills, know-how and know-why through rules of the craft and traditions of apprenticeship)

Such systems may require a higher initial investment but provide options for long duration. The review of the literature on the stability of systems and resilient (socio-ecological) systems displays a number of similar properties (Madni and Jackson, 2009). Due to their double nature as an attribute and a process, the principles of resilience, stability and robustness are generally formulated as heuristics.

Constructions or systems that are prone to be vulnerable typically result from

- Generalising experimental approaches (each building is a prototype)
- Use of 'new' materials without experience and sufficient evidence
- Change of programme and short-term decisions during the design and construction phase
- Complicated technical equipment systems (with incomplete documentation)

The principles of sustainable development reactivated the discussion on the relation between traditional cultural techniques for using resources economically and the discussion of the 'carrying capacity' of ecosystems. The concept of 'sustainable development' implicitly re-established a perspective that spans generations. However, those discussions have highlighted the difficulties of mapping developments of complex (cultural), social and ecological systems. After an initial, promising phase, the potential of sustainable development as a solution has become more focused on threats and risks. Beck (1992) refers to this phenomenon as the transition from a (modern) industrialised society to a 'risk society'. The returns from the growing system performance and the related complexity are no more the source of (equal) distribution of the fruits of progress but the source of (unequal) distribution of risk (Beck, 1992). Lovins and Lovins (1982) expressed similar doubts concerning the risk resulting from growing energy consumption. 'Economic, social and environmental threats can be seen as the dark side or shadow threat behind any vision of sustainability' (Moffatt, 2007, p. 158). The 'dark side' of this system thinking is the implied risk. In other words, it is very difficult to see the big picture when balancing the different needs to sustain social, economic, ecological and cultural capitals. Such properties can in the long run only be 'bought' with higher investments in options for the future. This is perhaps why there is a growing interest in the phenomenon of demolition and its cultural causes. Demolition is well known as omega-phase in the adaptive cycle and panarchy model (Gunderson and Holling, 2001) but it is much less understood as an urban transformation process. Little statistical data exist on demolition and the survival probabilities of building stocks are not well known (Bradley and Kohler, 2007). There are multiple reasons why buildings are demolished, deconstructed or destroyed (Thomsen et al., 2011) but no comprehensive model has been established. The idea to look to the resilience concept may provide a promising underpinning for such a model.

In conclusion, artefacts that have existed over a long period of time can be considered examples of resilience. They have survived a series of unpredictable disruptions during their lifetime. Their makers did not know the type and the dimension of the impacts. The built environment as such is a long-term document of adaptation and of how catastrophes have been overcome in history. The lessons from the survival of the built environment (individual buildings, as well as larger areas, i.e. building stocks, neighbourhoods and cities) can be used to inform, test and adapt the notion of resilience.

2.5 'FAST' AND 'SLOW' RISKS, PLANNING AND ANTICIPATION

The aim and inherent paradox of twentieth-century modernism within the building industry were the programmatic reduction of craftsmanship complexity by an industrialised normalised production approach. Industrial production methods became the standard in building and construction. This allowed for faster construction and fast growth, but at the cost of a shortened lifespan of the artefacts. Transmitted notions of quality were abandoned in favour of 'guaranteed performance' determined by specific, narrow programmatic definitions (Ellingham and Fawcett, 2006). Within the real estate sector, the expected return on investment was increasingly adapted to general (financial) market conditions. In this sense, cultural goods have also become subjected to the rules of the market. In this case, scarcity produces desirability; aged artefacts cannot be reproduced and thus have a rarity value. This pertains to a very small sub-stock of real estate. The results of this trend became visible at the onset of the twenty-first century. Central locations of surviving European cities are subject to modern real estate ventures with consequences that lead to their demolition and replacement. Cultural capitals that have been collectively produced over a long period of time are 'commodified' (Rifkin, 2001) and pass into private use.

The regime of modern planning followed the principles of anticipation of a set of particular, desired vision(s) of the future. Conservation existed only on the fringe of mainstream modernist thinking (e.g. heritage protection). By and large, the agenda was set to 'replace' the old with the 'new'. The neglect of very long time constants and the slow dynamics of the 'built environment' tend, however, to distort the understanding of the system itself and to underestimate the long-term risk consequences. In addition, risks in the industrialised world are highly diverse and a comparative evaluation leads to systemic problems. Each factor or combinations of factors reduce the understanding of general long-term developments. Complex, nested systems and various socio-cultural processes that constantly change what is deemed worthy of 'appreciation' exacerbate the difficulty of understanding loss risks, for example, 'multi-scale reciprocity' (Gunderson and Holling, 2001, p. 98).

The built environment features both 'slow-moving risks' and 'fast-moving risks', which have different risk profiles. It is especially challenging to anticipate the consequences of slow-moving risks such as the proliferation of additives that have a long-lasting toxic effect. In the field of cultural heritage protection, disaster prevention is traditionally realised through 'passive' measures. These can take the form of documentation, storage, protection through structural measures, education and training. In principle, it is impossible to reproduce aged artefacts that are of artistic or scientific importance. The practice of transferring the consequences of 'big risks' to society itself (e.g. political bodies act as an insurer of the last resort creating a 'self-insurance carrier') follows this insight.

Although current discourses consider the vulnerability of systems of the built environment, a differentiation between long-term risk consequences and possible short-term losses is still missing. A new debate on resilience needs to examine how to differentiate the various risk dynamics, but also must include a discussion on the quality of the losses. The simple ability of a system to adapt to new conditions, or to absorb shocks after a disruption, does not provide any information on the quality of the subsequent situation. For example, a society that loses large parts of its cultural heritage could find different stable 'alternative solutions'. However, these solutions cannot actually substitute the qualities that have been lost. The discussion on the description, qualification and preservation of cultural capitals has therefore formulated approaches which may be fruitful to a theory that includes questions of quality. Some examples are discussed below.

2.6 CULTURAL, NATURAL AND HUMAN-MADE CAPITAL

There are multiple definitions of cultural and of natural capital. However, these definitions are strongly interrelated (Berkes and Folke, 1994). Natural capital consists of non-renewable resources, renewable resources and environmental services. Cultural capital refers to factors that provide human societies with the means and adaptations to deal with the natural environment and actively modify it (Berkes and Folke, 1994, p. 130). Different societies have developed different ways to deal with natural capital, indeed the conceptualisation of 'nature' is a culturally specific phenomenon:

The diversity of ways to deal with the [natural] environment is a significant part of cultural capital, and perhaps as important to conserve as biological diversity (Gadgil, 1987, quoted in Berkes and Folke, 1994, p. 130).

The fundamental interrelation between natural, cultural and man-made capital is that cultural capital is the interface between natural and man-made capital. A representation of these linkages and their feedback loops is shown in Figure 2.1.

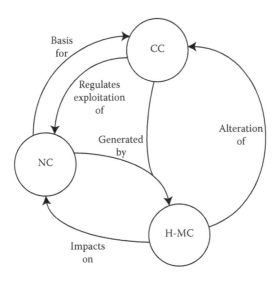

FIGURE 2.1   First-order interrelationships among natural (NC), human-made (H-MC) and cultural (C) capital. (Adapted from Berkes, F. and Folke, C. 1994. *Investing in Natural Capital*. Stockholm: Beijer International Institute, pp. 128–149.)

Natural capital is the basis, the precondition, for cultural capital. Human-made capital is created by the interaction between natural and cultural capital. Human-made capital, in turn may cause an alteration of cultural capital. Technologies which mask the dependence of societies on natural capital encourage the people to think that they are above nature (Berkes and Folke, p. 132).

The position of cultural capital is central but it is never neutral. Through institutions and institutional regimes, cultural capital may also lead to a more unsustainable use of natural capital and to the reduction of the carrying capacity. It may also lead to an increase of low-quality, human-made capital that will in turn result in losses of natural capital and cultural capital. Resilience would in this sense allow for positive feedbacks of cultural and natural capital and a growing value of the human-made capital.

2.7   CULTURE AS INTANGIBLE CAPITAL

Cultural capital comprises both intangible and material aspects (Throsby, 1995). In most cases it is the physical capital that comes to mind first. In the built environment this is capital that is objectified and embedded in objects, building, infrastructures and landscapes. Material cultural capital is continuously reduced through its (over-) utilisation and is exposed to normal ageing processes. Intangible capital consists of ideas, beliefs, collective memory, knowledge and ways of working/doing, but also art, music or literature. As a general rule, intangible capital is passed on from one generation to the next. This capital also needs to be maintained and can be increased with financial investments. Basically it is not subjected to wear (except 'oblivion') and can be used at will (Throsby, 1997). Intangible cultural capital, on the other hand, also contains knowledge and skills to maintain material cultural capital and thus contributes to the resilience of the system as a whole. The main threat to intangible cultural capital is the loss of knowledge relating to the built environment (both know-how and know-why).

The description of cultural capitals needs a frame of reference that allows for differentiation and value setting. Financial valuations can only reflect specific historic market conditions and have limited uses. If intangible values are linked with material artefacts and if intangible capitals interact with systems of the built environment, it is very difficult to assess how transfer, propagation and reconstruction are endangered.[*]

The cultural dimension in risk prevention therefore needs to aim at an 'abstract' level when information on 'new forms of risks' is central.[†] In this case, anticipation does not mean that known risk patterns are expected to be repeated (which can be

---

[*] Thorsby (2006) has analysed in detail the relations between cultural and economic capital: 'we can define an item of cultural capital as being an asset which embodies or yields cultural value in addition to whatever economic value it embodies or yields. The phrase "embodies or yields" is used here to emphasise the distinction between the capital stock and the flow of capital services to which that stock gives rise, a distinction which is fundamental to analysis of any sort of capital in economics … these flows also generate both economic and cultural value, which can, in principle at least, be identified and measured' (p. 40).

[†] Ulrich Beck by reference to Pierre Bourdieu's notions of 'reflexivity' (Beck, 1992, p. 221).

predicted based on existing information). Rather, anticipation means a systematic but 'non-linear' enhancement of knowledge to counter unknown hazards and paradoxical, new risk conflicts. The research of a stable or 'steady-state equilibrium' is therefore a possibly problematic (and simplified) metaphor for the future protection of the built environment and of cultural capitals. Examples of other and more promising directions might be investments in complexity, investments in 'purposeless' knowledge transfer, adaptation in the area of 'intangible' and cultural systems, foresight and anticipation for the future of material artefacts. They could be summed up under the concept of cultural resilience.

2.8 'NON-RECOVERABILITY' OF CULTURAL CAPITAL

While economic, physical, social and human capitals can at least partially be renewed and recovered, natural and cultural capital cannot be substituted in the same way. These capitals are only 'growing' slowly and develop only over a long period of time. It is proposed that the stock of a built environment can be positively characterised as cultural capital by

- Artefacts that have the potential for long-term survival.
- The possibility that knowledge and cultural techniques can be indirectly transmitted; this transmission depends upon the technical, artistic and artisanal qualities in the objects themselves.
- The multitude, number and size of the buildings.
- A broad distribution of objects regarding their age and variety.
- The ability that parts of the system are partially autarkic and can be independently used from other parts.
- The possibility of adaptation while existing parts can be further used to a large extent.

Cultural capitals are simultaneously material (tangible) and intangible: in many areas the 'knowledge archives'[*] of modern societies are linked to the stock of the built environment. The 'symbolic orders' (Bublitz, 1999, p. 77) of a society are structuring principles that are not necessarily bound to the built substance but which are historically represented through artefacts. Knowledge structures of cultures are reflected in historical practices and thus in the stock of the built environment. That is why 'resilience as a guiding principle' depends upon the following:

- The renewal of traditional mechanisms that transmit practices to preserve social values from generation to generation.
- A property (real estate) management approach that aims at risk prevention and has a long time horizon. In practical terms, this means a shift away from speculative short-term returns and anticipating possible future developments (as real options).

---

[*] The wording follows Bublitz (1999).

- A slower rate of usage, systematic maintenance and the provision of reserves are needed in particular situations with high uncertainty and/or long time spans, where anticipation and planning have little or no effect (Wildavsky, 1988).
- Functional and physical redundancy.
- Protection mechanisms for the transmission of 'non-recoverable capitals'. The survival of systems of great cultural and historical diversity entails more than a 'reactive' and object-related system for the listing of important monuments. It requires an 'open system' of a 'culture of stewardship'. A broad tradition, a tradition of knowledge and the history of ideas – all these factors serve as cultural techniques that also serve as indispensable resources.

A multitude of evidence proves that successful systemic decisions and developments of the built environment occurred over the past centuries. The idea and arrangement of a European city and its relevant sub-stock of surviving objects have, for example, survived continuously changing climates, use patterns and energy systems. Several sets of infrastructures (harbour installations, water engineering systems and streets) employ patterns that hardly ever need to be changed. The transmitted systems of traditional construction and preservation traditionally show high qualities of duration. However, more recent structures have higher 'new risk profiles' (i.e. loss of resilience due to reduced lifespan, a high degree of specialisation and the customisation of solutions and high connectivity of technical systems).

Potential hazards, in particular slow-moving hazards that are difficult to anticipate, concern other dimensions such as

- Complexity as risk (in the built environment there are the immeasurable interactions of system components, multidimensionality and 'behaviour' of sub-systems, the immensity of the complex knowledge that is needed for the system)
- Technological risk consequences and the risk of the development of technology itself
- 'New media' that standardise entire knowledge domains
- The loss of information and knowledge over a long time duration

The built environment is a 'depository of values' and is still subjected to comparably slow changes. Its ability to adapt slowly over time is well proven. The natural and cultural capital (of the built environment) can be conveyed to future generations if it is conservatively used and if adaptation occurs more slowly. The 'non-availability' of cultural capitals, however, demands modified strategies.

2.9   CONCLUSIONS

An examination of the concept of resilience revealed that it can provide remarkable complementarity to other concepts used in the sustainable management of buildings and infrastructures. The objective of continuity refers to stability and different types of equilibria. The ecosystem definition of resilience provides the possibility for the establishment of continuity through different types of equilibria. Instead of a rigid form of stability, which is only possible for short periods, continuity integrates controlled change and long-term adaptation. The essential criterion is the speed of transformation. Transformations that occur too quickly lead to a loss of values through overshoot; those that occur too slowly may lead to dereliction (Kohler and Hassler, 2003). The distinction between stable equilibrium (engineering stability) and multiple equilibria (resilience) can be found also in the distinction between the short-term behaviour of buildings and infrastructures (stability) and the long-term survival of the urban fabric through a slow adaptation (learning) process.

The incorporation of the resilience concept into the sustainable management of the built environment implies both long foresight and a differentiation between time constants, scales and dimensions (capitals). Different scales have clearly different time constants and it is possible to associate the different dimensions, in the sense of capitals, to the time-scale categories. As a design principle, resilience increases according to the expectations for time scale (longevity) and can be used as a central timing and memory concept. On the other hand, anticipation can only refer to shorter time frames and the corresponding scales and dimensions. Anticipation is therefore a strategy that allows a fast reaction and learning process and is central to risk management (Figure 2.2).

The reduction of vulnerability and the increase of resilience depend upon an increased control over time in conceptual and practical ways: time horizons, scenarios, future options, etc. In this interpretation resilience would be a superior guiding principle that harnesses the lessons from long-term surviving systems but that also encompasses the limited ability to anticipate the future and make predictions. Resilience would therefore become a central 'timing tool'.

The remarkable robustness of many historical structures and at the same time their capabilities to adapt to a changing environment constitute positive examples

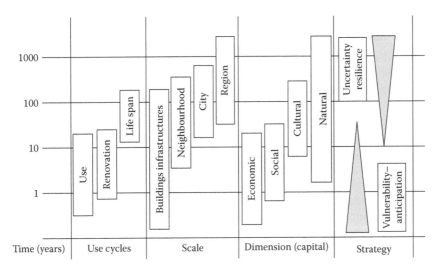

FIGURE 2.2   Time constants, scales, dimensions and long-term strategies for the built environment.

of resilience. The built environment can be understood as a long-term document of adaptation and of how catastrophes have been overcome in history. It implicitly contains lessons for resilience.

The built environment is exposed to both slow-moving risks and fast-moving risks, which have different risk profiles and therefore need different approaches. This would consist in differentiating risk dynamics but it also needs to articulate the quality of the losses. Although a system may have some adaptive capacities, this does not guarantee the quality of the subsequent situation.

Cultural and natural capitals were introduced as a crucial aspect of resilience in the sense that they cannot be substituted, because they cannot be reproduced. Their position is 'critical'. Resilience would in this sense allow positive feedback loops for cultural and natural capital and a growing value of the human-made capital. Cultural capital (both in material and intangible forms) promotes memory (slow cycles), understanding and learning (fast cycles) and anticipation (virtual cycles and changes of scales).

The criticality or 'non-availability' of cultural capitals is not a given as such; it demands modified strategies. It will be necessary to invest in material as well as in intangible cultural assets and to slow the rate of change in the built environment. The traditional over-investments in resilient structures were relatively simple before the twentieth century. The redundancy of parts, the reparability of components, the diversity of artefacts, the width of the domain of possible solutions, the limitation of growth and the maintenance of traditional skills were common methods. The repetition of these 'successful' approaches in the anticipation of future risks can possibly be at least partially successful if these risks are known. In the societal sphere as well as in the realms of colonised nature many of the future interactions and the probability of survival of particular structures are not understood or are not yet understandable. In the built environment the reasons for disappearance are often not visible. The repetition of the concepts that were successful for survival in the past will probably not be successful in the future. It will be necessary to invest in material as well as in intangible cultural assets and it will be necessary to slow down the use of the built environment. The basic distinction of Wildavsky (1988) between vulnerability and anticipation versus uncertainty and resilience is fundamental. Resilience can be operationalised (measured) when referring to certain types of clearly defined ecological and social systems. But for more complex systems such as the built environment there are no simple models and certainly no simple metrics available. A strategy based on anticipation is possible for more or less known threats, essentially by reducing vulnerability and increasing adaptive capacity. Coping with uncertainty due to unknown threats, unknown combination of threats, unknown reactions of the built environment to natural and man-made disruptions can only rely on heuristics derived from the observation of successful outcomes. Resilience in this sense has little to do with the exact description of an ecological or social system; it is more a design rule for complex systems. It is no more a descriptive concept but a normative concept (Brand and Jax, 2007). This does not diminish the value of the transdisciplinary discussion that has developed around resilience, a discussion that has already produced a number of new insights and will certainly continue to do so.

REFERENCES

Assmann, A. 1999. Zeit und Tradition. Kulturelle Strategien der Dauer. Köln.

Assmann, J. 1999. Das kulturelle Geda¨ chtnis, Schrift, Erinnerung und politische Identität in frü hen Hochkulturen.

Assmann, J. 2010. *Cultural Memory and Early Civilization: Writing, Remembrance, and Political Imagination*. Cambridge: Cambridge University Press.

Beck, U. 1992. *Risk Society: Towards a New Modernity*. Munich: Sage.

Berkes, F. and Folke, C. 1994. Investing in cultural capital for sustainable use of natural capital. In: A. M. Jansson, M. Hammer, C. Folke and R. Costanza (eds.). *Investing in Natural Capital: The Ecological Economics Approach to Sustainability*. Washington, DC: Island Press. pp. 128–149.

Bradley, P. and Kohler, N. 2007. Methodology for the survival analysis of urban building stocks. *Building Research and Information*, 35(5), 529–542.

Brand, F. S. and Jax, K. 2007. Focusing the meaning(s) of resilience: Resilience as a descriptive concept and a boundary object. *Ecology and Society*, 12(1): 23 [online]: http://www.ecologyandsociety.org/vol12/iss1/art23.

Bublitz, H. 1999. Foucaults Archä ologie des kulturellen Unbe- wussten. Zum Wissensarchiv und Wissensbegehren moderner Gesellschaften. Frankfurt/New York: Campus-Verlag.

Elias, N. 1992. *Time: An Essay*. Oxford: Blackwell.

Ellingham, I. and Fawcett, W. 2006. *New Generation Whole-Life Costing: Property and Construction Decision-Making under Uncertainty*. London: Taylor & Francis.

Fischer-Kowalski, M. and Weisz, H. 1999. Society as hybrid between material and symbolic realms: Towards a theoretical framework of society–nature interaction. *Advances in Human Ecology*, 8, 215–251.

Folke, C. 2006. Resilience: The emergence of a perspective for social–ecological systems analysis. *Global Environmental Change*, 16, 253–267.

Gunderson, L. and Holling, C. S. 2001. *Panarchy: Understanding Transformation in Human and Natural Systems*. Washington, DC: Island Press.

Hassler, U. and Kohler, N. 2004. *Das Verschwinden der Bauten* des Industriezeitalters. Tübingen/Berlin: Wasmuth Verlag.

Hoffenson, S., Dagman, A. and Söderberg, R. 2013. A multi-objective tolerance optimization approach for economic, ecological, and social sustainability. In: Nee, A. Y. C., Song, B. and Ong, S. K. *Re-engineering Manufacturing for Sustainability*. Singapore: Springer. pp. 729–734.

Holling, C. S. 1996. Engineering resilience versus ecological resilience. In: Schulze, P. (Ed.), *Engineering within Ecological Constraints*. National Academies Press. p. 31–44.

Hollnagel, E., Woods, D. D. and Leveson, N. 2007. *Resilience Engineering: Concepts and Precepts*. Farnham: Ashgate Publishing, Ltd.

Kohler, N. and Hassler, U. 2003. Integrated life cycle analysis in the sustainability assessment of urban historical areas. SUIT www.lema.suit.be

Küttler, W., Jörn Rüsen, J. and Schulin, E. (Eds.). 1994. Ein Überblick zu den geschichtstheoretischen Konzepten. In: *Geschichtsdiskurs Band 2, Anfänge des modernen historischen*. Frankfurt: Denkens Fischer Verlag.

Lawrence, D. L. and Low, S. M. 1990. The built environment and spatial form. *Annual Review of Anthropology*, 19, 453–505.

Lévi-Strauss, C. 1966. *The Savage Mind*. Chicago: The University of Chicago Press.

Lovins, A. B. and Lovins, H. L. 1982. *Brittle Power: Energy Strategy for National Security*. Andover, MA: Brick House Publishing Company.

Madni, A. and Jackson, S. 2009. Towards a conceptual framework for resilience engineering. *IEEE Systems Journal*, 3(2), 181–191.

Mileti, D. S. 1999. *Disasters by Design: A Reassessment of Natural Hazards in the United States*. Washington, DC: The Joseph Henry Press, National Academy of Sciences.

Moffatt, S. 2007. Time scales for sustainable urban system design. Dissertation. University of Karlsruhe, Karlsruhe.

Moffatt, S. and Kohler, N. 2008. Conceptualizing the built environment as a social–ecological system. *Building Research and Information*, 36(3), 248–268.

Rapoport, A. (Ed.). 1976. *The Mutual Interaction of People and Their Built Environment: A Cross-Cultural Perspective*. Chicago: Aldine, pp. 7–35.

Rapp, F. 1992. *Fortschritt, Entwicklung und Sinngehalt einer philosophischen Idee*. Hamburg: Darmstadt.

Rapp, F. 1994. *Die Dynamik der modernen Welt. Eine Ein-* führung in die Technikphilosophie. Hamburg: Junius.

Reichholf, J.H. 2008. *Stabile Ungleichgewichte*. Die Ökologie *der Zukunft*. Frankfurt: Suhrkamp Verlag.

Rifkin, J. 2001. *The Age of Access*. New York: Penguin.

Sauer, C. 1925. The morphology of landscape. *University of California Publications in Geography*, 22: 19–53.

Thomsen, A., Schultmann, F. and Kohler, N. 2011. Deconstruction, demolition, destruction (special issue). *Building Research and Information*, 39(4), 327–429.

Throsby, D. 1995. Culture, economics and sustainability. *Journal of Cultural Economics*, 19, 199–206.

Throsby, D. 1997. Sustainability and culture some theoretical issues. *International Journal of Cultural Policy*, 4(1), 7–19.

Thorsby, D. 2006. The value of cultural heritage: What can economics tell us. In: Clark, K. *Capturing the Public Value of Heritage: The Proceedings of the London Conference 25–26, January 2006*. Swindon, England: English Heritage.

Wildavsky, A. 1988. *Searching for Safety*, Vol. 10. New Brunswick, NJ: Transaction Publishers.

3 Processes of Resilience Development in a Population of Humanitarian Aid Workers

Amanda Comoretto

CONTENTS

3.1 INTRODUCTION

The bamboo that bends is stronger than the oak that resists.

(Japanese proverb)

Resilience is a term extensively used to describe the factors that promote well-being and strength in individuals who are undergoing unusually stressful life conditions. It is an interactive construct concerned with acutely traumatic experiences followed by positive psychological outcomes despite those experiences (Luthar, 2003). In an attempt to better understand this construct, the present chapter will first review commonly understood definitions of the term, as well as key protective factors related to the development of resilience. Against this background, the second part will evaluate resilience research in populations of aid workers, stress factors associated with their job and consequences of experienced trauma. In the final part of the chapter an investigation conducted between 2004 and 2008 at London South Bank University by Comoretto et al. (2011), aimed at investigating processes of resilience development in a group of humanitarian aid workers, will be discussed. The theoretical model developed by the authors of this study to account for the development of resilience will be reviewed together with the devised method to test such a model as well as with the results of the investigation. Moreover, possible explanations for the development of resilience will be proposed. It will also be suggested that this study contributed to the advancement of theory within the area of resilience inquiry.

3.2   DEFINING THE CONSTRUCT OF RESILIENCE

The term 'resilience' entered the field of psychology from applied physics and engineering, where it was originally referred to the property of some materials to bounce back and resume their original shape or condition. In information technology (IT) resilience concerns the capacity of a system to function despite anomalies. Therefore, this construct cannot be reduced to simple resistance to hostile agents because this would imply rigidity. It rather reminds us of qualities such as adaptation and elasticity. In addition, it is not simply equated to invulnerability (Anthony, 1982) because this would imply resistance to shock coupled with paralysis of the individual and would therefore be very far from the actual meaning of the term. In 1989 Professor Michael Rutter, probably the most famous scholar of resilience, described this construct as a phenomenon 'shown by the people who do well despite having experienced a form of stress which, at least in the population as a whole, is known to carry a substantial risk of an adverse outcome' (p. 23). Some years later the American psychologist Ann Masten (1994) tried to explain successful adaptation despite risk and adversities, suggesting that individuals can bend, yet subsequently recover. In order for this to happen, the person must be doing well. Fonagy et al. (1994) considered resilience as a normal developmental process despite difficult external circumstances. The French scholar Cyrulnik (1999) reminded us that psychological resilience can be defined as the ability to live and develop positively, in a socially acceptable way, despite stress or adverse conditions bringing about the risk of negative outcomes. In 2001 Masten postulated that resilience manifested itself in individuals' daily behaviours and life patterns and could be ascribed to three kinds of phenomena: (1) good outcomes despite a high-risk status, (2) sustained competence under threat and (3) recovery from trauma. In summary, over the years resilience has been viewed by many as the ability to live in a satisfactory way despite hardship (de Tychey, 2001).

Research focusing on individuals exposed to biological risk factors and stressful life events has gone through several stages in an attempt to understand vulnerability and resilience. At the beginning of the 1980s, emphasis was placed on the negative developmental outcomes associated with a single risk factor, such as low birth weight, or with stressful life events, such as the prolonged absence of a parent following divorce. Investigators were concerned with the exploration of possible correlations between stressful life experiences (risk factors) and a range of psychiatric disorders. Risk was used to refer to variables that increased individuals' likelihood of psychopathology or susceptibility to negative developmental outcomes. Some risks were deemed to be found internally, resulting from the unique combination of characteristics that made up an individual, such as temperament or neurological structure. It was thought that other risks could come from the external environment, namely from all those events beyond the normal range of human control, such as accidents, family re-arrangement, poverty, war, armed conflict, mass murder, famine and mass displacement.

The second phase of resilience research was marked by the identification of specific personal traits or protective factors that allowed people to recover from adversity (see Section 3.3.1). These appeared to transcend ethnicity, social class and geographic boundaries; they were also thought to occur together within a particular developmental period or within a particular population (Gore and Eckenrode, 1994). Moreover, their effects were only visible in their interaction with risk. Additionally, protective factors were thought to shape, to a large extent, the strategies that individuals used to manage stressful situations and to defend themselves against painful experiences.

The most recent wave of resilience inquiry, in the past 15 years or so, has been an attempt to answer the question 'How are resilient qualities acquired?' With the expansion of the literature on this topic it became clear that individual and environmental factors were necessary but not sufficient to understand the construct of resilience. As a consequence, theoretical frameworks of resilience development emerged in the literature. Person-focused approaches identified resilient people whose lives and attributes were studied by investigators in comparison to maladaptive individuals with similar levels of risk or adversity displaying markedly different outcomes. Moreover, variable-focused approaches tested for links among measures of the degree of risk, outcome and protective factors.

3.2.1 PROTECTIVE FACTORS IDENTIFIED IN RESILIENCE RESEARCH

As seen before, specific personal qualities may influence or modify responses to adversity. The objective of this section is to provide a description of the main protective factors and evidence for their effectiveness in changing responses to stressful life events. When possible, evidence drawn from studies conducted with humanitarian aid workers will be presented.

The most important protective quality can be considered normal cognitive development, which in the literature on resilience has emerged as a key factor in many forms, including average or better IQ scores and good attention skills (Masten et al., 1988; Seifer et al., 1992; Herrenkohl et al., 1994; Fergusson and Lynskey, 1996). Research studies have underlined how a high IQ can play a key moderating and protective role in the process linking adversity to social conduct (Kandel et al., 1988; White et al., 1989; Masten et al., 1999). According to Block and Kremen (1996) the capacity for cognitive problem-solving behaviour has evolved as the most powerful means of affective adaptation but 'it is adaptation that is the key, and intelligence is but one way' (p. 359).

In the literature on at-risk children and adolescents there have been a number of suggestions stressing the fact that gender may influence or modify response to adversity. Girls have often been described as being more resilient than boys in childhood (Werner and Smith, 1982; Rutter, 1989), whereas males have proved to be more emotionally vulnerable than females in childhood to the effects of biological insults, care giving deficits and economic hardship (Werner, 1993). This trend is thought to be reversed in the second decade of life, with girls becoming more vulnerable than boys in adolescence. In young adulthood the balance appears to shift back again in favour of women. Similarly, it appears that with age the psychological protective system for human development can increasingly come under the control of the individual (Scarr and McCartney, 1983). Older children and adolescents have often been reported to have stronger and longer lasting positive reactions to major stressors than very young children (Masten et al., 1990). According to Aguerre (2002) the very same definition of resilience in young children and adolescents, namely the ability to overcome risk and trauma thanks to their psychological resources, can potentially be applied to ageing people in the process of coping with their changing biological status. A good physical condition is usually predictive of resilience, too. Individuals with few physical problems, good sleeping patterns and physical strength may internalise this physical wellness and interpret themselves as strong psychologically (Kumpfer, 1999).

Length of work experience can be considered as a positive as well as a negative factor in developing resilience. In a survey conducted among Kosovar Albanian aid workers to detect rates of mental health problems (Cardozo and Salama, 2002), a greater proportion of individuals suffering from post-mission depression and anxiety were on their first mission and had worked for their organisation for less than 6 months. Conversely, the Centre for Disease Control mental health survey in Kosovo (Cardozo et al., 2005) found that workers who had more than five previous assignments, and were therefore seen as being quite experienced, were at significantly higher risk for developing stress-related problems than those with only two or three career assignments. The authors of the survey suggested that 'mental health problems may be correlated with the number of trauma events experienced, which are, in turn, related to number of missions and amount of time spent in the field' (p. 252). As a result, extremely experienced aid workers, as well as first-time staff, could be at risk for negative mental health outcomes. On a similar note, a survey by Corneil and colleagues (1999) found that fire fighters with more than 15 years of service had a significantly increased odds ratio of post-traumatic stress disorder (PTSD) than their less experienced colleagues.

Social support refers to the existence or availability of people on whom one can rely. Research with adults (Conger et al., 1999) has shown that social support can protect people in crisis from a wide variety of pathological states: from low birth weight to death, from arthritis through tuberculosis to depression, alcoholism and substance dependence. Well-developed social skills have also been described as being positively correlated with resilience (Masten et al., 1990) and the resolution of traumatic experiences has often been found to be mediated by the availability of effective and meaningful social support. Procidano and Heller (1983) emphasised the role of social support in mediating the relationship between life events and distress. Social support can, in fact, contribute to positive adjustment and personal development, providing a buffer against the effects of trauma (Caplan, 1979). Kaspersen et al. (2003) compared social networks as moderators between trauma exposure and post-traumatic symptoms in two samples: relief workers and UN soldiers. The social network, measured as family, friends, neighbours and colleagues, was found to moderate the relation between trauma exposure and reactions. Among UN soldiers social support was important for those low on trauma exposure, while among relief workers support was important in the high-exposure condition. Conger et al. (1999) investigated how high marital support was likely to reduce the consequences of economic pressure in predisposing to emotional distress and found that effective couple problem-solving techniques reduced the adverse effect of marital conflict on marital distress. According to Cardozo and Salama (2002), 'family networks can be particularly important in offsetting stressors encountered by aid workers' (p. 250). To conclude, close connections with families and friends can potentially help humanitarian workers overcome the sense of culture shock experienced when re-integrating back home.

Humanitarian organisations have increasingly reported the successful use of peer support networks made up of volunteer staff who receive appropriate training and operate under the supervision of professionally trained counsellors. The value of peer support is said to lie in the possibility to share similar experiences. Research findings indicate that it is very important to talk to others who empathise, understand and put one's feelings into perspective (Stewart and Hodgkinson, 1994). Some studies (Gibbs et al., 1993; Sarason et al., 1995) found that individuals who perceived themselves as being a part of a social and professional network were less negatively affected by stressful events and also less likely to experience stress-related problems. Similarly to peer support, organisational support can be critical in highly demanding occupations. In the military setting intensive training and tight-knit groups are protective factors for soldiers, allowing them to be better prepared to function amid violent conflicts. The extent to which this military experience can be extrapolated to the civilian domain is still, however, rather unclear (Cardozo and Salama, 2002). Eriksson et al. (2001) carried out a survey involving 113 recently returned staff from humanitarian aid agencies. Participants completed questionnaires pertaining to their exposure to traumas while overseas and to their emotional responses during the first 6 months of re-entry. Those with high levels of exposure reported a lower number of PTSD symptoms when they indicated highly perceived organisational support, suggesting the importance of this component in buffering the effects of stress.

Socio-cognitive abilities have also been found to be related to resilience (Masten et al., 1990). The ones most commonly reported in the literature are coping strategies and motivation.

3.2.1.1 Coping Strategies

Coping can be described as the process by which one manages the demands and emotions generated by that which is appraised as stressful (Lazarus, 1999) or, in other words, 'the transactional process between individuals and their environment, involving appraisals such as whether the situation or event is a threat, a challenge, or a loss, and appraisals of what can be done' (Park, 1998, p. 271). Coping can be broadly differentiated into efforts to change the environment, and efforts directed inwards to change the meaning attached to an event and to increase understanding. Working in international development and emergency aid requires the improvement of emotional strategies and technical skills necessary to cope with various situations. Seyle (1985) has suggested that coping strategies differ in effectiveness, and that unproductive or inappropriate approaches may promote the detrimental effects of stress. For example, some coping strategies such as overworking may be counterproductive. Because of feelings of guilt about their limited effectiveness, humanitarian aid workers often attempt to overcome these emotions through denial and immersion into work, something that enables them to withdraw emotionally from their dilemma. This intense involvement can, however, lead to exhaustion and burnout (Paton, 1992; Straker, 1993).

Distancing or detaching from a stressful situation is also depicted as a major coping strategy in the literature on humanitarian staff (Gibbs et al., 1993; Mitchell and Dyregrov, 1993). This mechanism enables workers to cope emotionally with their experiences and prevents them from becoming incapacitated by their reactions, thereby facilitating effective role performance (Mitchell and Dyregrov, 1993). Blame and denial are also commonly used. Aid workers may be able to redirect feelings of guilt and failure to alleviate human sufferings away from themselves. In this case they are likely to blame other factors such as governments, coordinators, bureaucracy, etc. so that failure to act is rationalised through claims of having no power to transform the inadequacies of the system. Whilst external constraints are indeed very real, workers are reluctant to admit the possibility of their actions contributing to failures (de Waal, 1988). Not infrequent also is the response of rationalising actions by holding convenient beliefs, which has the effect of dehumanising the population helped. For instance, commonly held beliefs are that traumatic events are not disturbing to refugees. 'Africans do not feel pain as we do. These people are always surrounded by death and sickness, so it's nothing new for them' (de Waal, 1988, p. 7).

3.2.1.2 Motivation

Human motivation is the drive that allows individuals to set control over their lives and to reach targeted objectives. Locus of control (LOC), optimism and self-efficacy beliefs can be considered the proximal determinants of human action, as well as the main components of motivation (Bandura, 1989).

LOC refers to one's personal belief system about the ability to affect an outcome. If the individual believes that outcomes occur by chance and that he has no influence over them, it is possible to talk about an external LOC. On the contrary, individuals exhibiting an internal LOC believe that events can be influenced and modified (Rotter, 1966). Resilient individuals are thought to be characterised by an internal LOC and are more hopeful about their ability to create positive outcomes for themselves and others (Luthar, 1991; Werner and Smith, 1992). In contrast to individuals with a pessimistic explanatory style, who perceive bad events as the result of internal, global and stable factors, those characterised by an optimistic explanatory style are more inclined to view negative experiences as a consequence of external, transient and specific factors. Speculating about prevention, Seligman (1991) suggested that optimism can be learnt early in life and that it helps to achieve more and maintain better health. The work of Albert Bandura (1977) focused on self-efficacy, defined as a belief in one's capabilities to organise and execute the course of action required to attain a goal. Efficacy is linked to the terms *effective, efficacious and control.* Self is generally defined as the identity of a person. Therefore, the definition of self-efficacy implies a conscious awareness of one's ability to be effective, to control actions or outcomes. People's beliefs in their ability to solve problems are positively related to the likelihood of initiating instrumental actions to reach targeted goals.

Having explored and described some key concepts related to the historical development of the construct of resilience, as well as the main protective factors identified as being responsible for its development, the next part of this chapter will focus on research into humanitarian aid workers, stress factors associated with their job and processes of resilience development.

3.3   STRESS, TRAUMA AND AID WORKERS

People who work in international development and emergency aid represent a multiplicity of different professionals, mainly health personnel and logisticians, but also engineers in the case of natural disasters. To date, very few studies have investigated the background characteristics of this group. Little is known about them except for the over-idealised stereotypical image dominated by ideas of heroes fighting against injustice and suffering (Beigbeder, 1991). Aid workers are commonly cast in the role of helpers, and a commitment to this kind of work is considered a heroic vocation, conferring an attractive aura of adventure and mystery to those who dedicate themselves to it.

It has been suggested (Raphael, 1984; Stearns, 1993) that aid workers represent hidden victims of their profession because of the potential powerlessness of their position, which may put them at risk of emotional and psychological problems. The literature on this topic points out that workers' motivation can switch from the genuine desire to help others to diverse complex and multifaceted phenomena (Beigbeder, 1991). Nonetheless, despite the informal recognition of a range of motivations that move these people, there is a tendency to regard their reasons as primarily altruistic and noble (Fox, 1995). The stereotyped helper image continues to be upheld by organisations and workers alike. Summerfield (1990) partly relates this trend to the culture of the agency, which tends to view staff members as perfectly able to cope with any given situation. Raphael et al. (1986) suggest that workers themselves may consider it inappropriate or unprofessional to feel and show emotional vulnerability.

Working in the midst of uncontrollable disasters may give rise to negative emotional reactions, especially since experienced and effective workers are sent consecutively from one disaster zone to another (Austin and Beyer, 1984; Jones and Jones, 1994). In addition, exposure to chronic high levels of stress is amplified by residing in insecure environments. While those attracted to this job may be statistically more emotionally stable than the general population and less likely than the ordinary citizen to collapse under intense pressure, aid workers are nevertheless affected by their daily tasks and may become the unrecognised victims of their profession (Raphael et al., 1986). As early as 1913 an article published in the *British Medical Journal*, based on a study of 1479 missionaries, reported that nervous illnesses of a neurasthenic type (characterised by fatigue) were the most common cause of premature repatriation (Price, 1913). Macnair (1995) found that 12% of field staff returning from difficult missions showed some form of psychological distress. Donovan (1992) highlighted how 25% of aid workers interviewed in his study had returned home prematurely, whereas about 50% had kept on working with reduced efficiency because of experienced stress.

Although transient psychological difficulties are common following exposure to stressful situations (Durham et al., 1985; Lane, 1994), enduring psychiatric long-term problems can become the norm. Among the most commonly reported mental health issues it is possible to find vicarious traumatisation, defined as 'a group of disruptive and painful psychological symptoms that result from exposure to other people's trauma' (McLean et al., 2003, p. 417). Burnout is also a central phenomenon related to trauma and can be described as 'a syndrome of physical and emotional exhaustion involving the development of a negative self-concept, negative job attitudes and loss of concern and feelings for the others' (Pines and Maslach, 1978, p. 233). Finally, the most stressful experiences are linked to one specific diagnostic category, namely PTSD, identified by the World Health Organization as the most severe psychiatric disorder resulting from a catastrophe (World Health Organization, 1992). PTSD, often a normal reaction to an abnormal event, is characterised by a combination of (a) intrusive symptoms linked to re-experiencing the traumatic event, (b) avoidance and numbing symptoms, such as avoiding activities, places, thoughts or feelings that remind of the trauma and (c) increased anxiety and emotional arousal (Horowitz, 1993). Lastly, earlier research identified depression as a significant mental health risk for aid workers (Richardson, 1992), whereas clinicians working with this population suggest that common reactions to missions' stress can also include loss of self-esteem, anger, psychosomatic problems and sexual promiscuity (Donovan, 1992; Carr, 1994).

3.3.1 STRESSORS AFFECTING HUMANITARIAN WORKERS

Stressors are those single, multiple or complex events that bring about emotional strain. Commonly reported difficulties during a mission include facing large-scale poverty, injustice, suffering, despair, death, powerlessness, overwhelming responsibility, ethical dilemmas, role ambiguity, communication problems, unpredictable circumstances, cross-cultural adjustment, isolation, etc. (Paton, 1992; Slim, 1995). Moreover, aid workers are at increased risk of illness and injury because of the logistic arrangements in the areas in which they work, as well as the inadequate medical facilities. Schouten and Borgdorff (1995) found that, compared with colleagues remaining at home, mortality rates were doubled among aid workers on a mission in their study. Furthermore, these individuals tend to have high ideals and expectations about being able to achieve great results, which may put them at particular risk of experiencing emotional exhaustion (Stearns, 1993). They also tend to be active for longer hours (Paton, 1992). In one study of aid staff work patterns, 50% of participants claimed they regularly worked more than 60 h a week (Macnair, 1995).

Compared to development work, complex humanitarian emergencies and natural disasters may generate more stress among workers for several reasons. First of all, an element of physical insecurity, with the risk of violent personal assault or injury, is constantly present. Second, working in complex emergencies or natural disasters necessarily involves moral and ethical dilemmas (e.g. witnessing human rights abuses, being constrained from responding by operational considerations and enduring concerns about humanitarian aid perpetuating conflict). Finally, caring for people with serious injuries caused by violence, witnessing unnatural deaths and handling dead bodies or body parts can be considered highly traumatic experiences in themselves.

3.4 A STUDY OF RESILIENCE IN HUMANITARIAN AID WORKERS

As mentioned in Section 3.1, this last section will focus on a study conducted between 2004 and 2008 at London South Bank University by Comoretto et al. (2011) and aimed at investigating how some of the protective factors described in Section 3.3.1, likely to influence the development of resilience, interacted in a group of humanitarian aid workers before and after deployment in the field, resulting in higher or lower levels of resilience.

In the last two decades, there has been an abundance of research on child resilience, whereas little has been done to examine the protective and vulnerability processes unique to adulthood, such as career changes, marriage, having children or building of social networks. When adult populations have been studied, the emphasis has prevalently been on health care workers (doctors and nurses), military personnel, war veterans, refugees, policemen, fire fighters and social workers. Only a few studies have been produced on the psychological difficulties encountered by humanitarian aid workers, reflecting a lack of awareness on the part of institutions of the distress that these people experience daily. The potential development of resilience processes in this population has also been overlooked. Unfortunately, because much of psychology's knowledge about how adults cope with loss or trauma comes from individuals seeking treatment or exhibiting great distress, trauma theorists have often viewed resilience as either rare or pathological.

In Section 3.3 the dynamic development of resilience was described as a system which can be learnt at any point in life and not just during childhood or adolescence, which leaves space for studies focusing on adult populations. Egeland et al. (1993) found that individuals who recovered more readily following periods of maladaptation had a history of adaptation over time and across phases of development. They concluded that 'rather than being a childhood given or a function of particular traits, the capacity for resilience develops over time in the context of environmental support' (p. 518).

Between 2004 and 2008 Comoretto and colleagues developed a theoretical model aimed at explaining positive or negative changes in resilience. Their hypothetical framework proposed that cognitive and environmental factors were likely to impact on an individuals' psychological responses during stressful life experiences. At the same time these factors were mediated by considered fixed dispositional markers, which in this study were gender, age, number of previous field missions, general health status, marital status and age at which participants had left education. Environmental protective markers included relationship networks: family members, work associates and friends. Considered cognitive protective factors were motivation, LOC and coping mechanisms. All the protective factors included were chosen after a critical assessment of research investigations conducted with adults involved in stressful and demanding jobs (policemen, soldiers, nurses, doctors, etc.).

The model was empirically tested using a longitudinal approach with measurements at two different time points, before and after a mission. This design allowed the examination of the impact of protective factors on behaviour adjustment before and after stressful experiences, such as those characterising humanitarian missions. The interrelationship of the three domains of protective features was predicted to impact on change in resilience, either positively by an increase in this construct or negatively by a decrease in the same. Moreover, the relationship between dispositional resources and outcome was hypothesised to mediate environmental stressors. To summarise, the model depicted in Figure 3.1 was tested in a humanitarian population to predict resilience and whether change in this construct would be affected by the identified protective factors.

In terms of methodology, a mixed-method investigation was carried out in two phases: a longitudinal self-completion questionnaire survey (phase I) and a series of semi-structured qualitative interviews (phase II). In phase I, questionnaires were completed by expatriate staff members of 36 humanitarian agencies before (time 1) and after (time 2) a field mission. A total of 56 people took part in the study, and approximately 51 returned the questionnaire at follow-up (91% of the cohort). The questionnaire incorporated previously validated scales to assess the various components of the devised theoretical model (see Appendix 3.1). Phase II involved the development of semi-structured interviews to investigate humanitarian workers' own accounts of field experiences. A schedule of questions was developed allowing for themes surrounding resilience to emerge, as well as to explore participants' reflections about field experiences. Participants' existing and changing social networks, as well as their importance in understanding the role of support in the experience of adversity, were explored, too.

Path analysis was the statistical technique used to test the model presented in Figure 3.1. People scoring high on the scale measuring PTSD, namely the Los Angeles Symptom Checklist (LASC), were more likely to report negative changes in

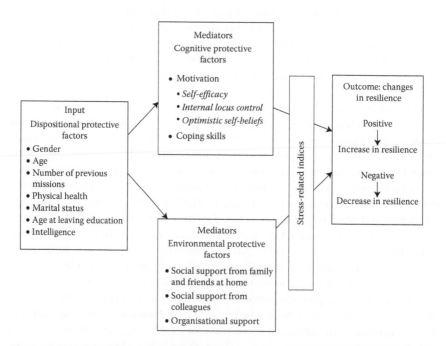

FIGURE 3.1 A model of dispositional, cognitive and environmental factors to examine changes in resilience in a population of humanitarian aid workers.

APPENDIX 3.1
Scales Included in the Final Questionnaire and Their Characteristics

Tool	Measured Construct	Part of the Theoretical Model Tested	N Items	Score Range	High Score (Highest Value)	Low Score (Lowest Value)
Los Angeles Symptom Checklist (LASC)	PTSD	Stress	43	0–4	More PTSD symptoms (172)	Less PTSD symptoms (0)
Maslach Burnout Inventory (MBI)	Burnout	Stress	22	0–6	Higher degrees of burnout (30)	Lower degrees of burnout (0)
COPE	Coping skills	Cognitive protective factors	60	1–4	More ability to put coping strategies in place (16)	Less ability to put coping strategies in place (4)
Generalised Self-Efficacy Scale (GSE)	Self-efficacy	Cognitive protective factors	10	1–4	High self-efficacy (40)	Low self-efficacy (10)
Adapted version of the Rotter's Internal-External LOC scale	LOC	Cognitive protective factors	10	0–1	External LOC (10)	Internal LOC (0)
Life Orientation Test (LOT)	Optimism	Cognitive protective factors	12	1–5	Less optimistic attitude (60)	More optimistic attitude (12)
Adapted version of the Social Provisions Scale (SPS)	Social support	Environmental protective factors	14	1–4	High availability of social support (56)	Low availability of social support (14)
Self-Report Questionnaire-20 items (SRQ-20)	General health	Dispositional protective factors	20	1–0	High number of symptoms present (20)	Low number of symptoms present (1)
Ego Resiliency Scale (ER-89)	Resilience	Resilience	14	1–4	High resilience (56)	Low resilience (14)
Connor–Davidson Resilience Scale (CD-RISC)	Resilience	Resilience	25	0–4	High resilience (100)	Low resilience (0)
Resilience Scale for Adults (RSA)	Resilience	Resilience	33	1–7	High resilience (231)	Low resilience (33)

resilience. Similarly, those participants employing mental disengagement as a coping technique, as measured by the COPE scale, were found to be characterised by negative changes in resilience at follow-up. Finally, those participants who actively looked for friends, family and colleagues in difficult times showed increased resilience compared to those who did not seek out available social support networks. The relationship that people who had left education at a younger age were more likely to experience negative change in resilience was largely dependent upon indirect paths via social support, the number of LASC symptoms reported, and the use of mental disengagement as a coping technique.

According to the results of qualitative interviews, motivations for starting up aid work (e.g. having strong humanitarian drives), deploying positive coping strategies (e.g. tolerance and cooperation), using mental preparation before a mission, and finally, mentally distancing from occupational stressors were cognitive protective factors helping participants minimise the effects of work-related stress and maximise positive changes in resilience. Quotations taken from the interviews exemplifying these key issues are given in Table 3.1.

Participants also detailed types of social support and the long-term consequences of humanitarian work experiences on pre-existing social relationships. Examples are shown in Table 3.2.

TABLE 3.1

Aid Workers Talking about Strong Humanitarian Drives, Showing Tolerance and Cooperation, Mentally Preparing for a Mission and Mentally Distancing from Occupational Stressors

A. Strong Humanitarian Drives

'I've always wanted to do human rights work, and then I wanted to do it in an international setting. This just sort of fit. This is what made me want to do that'.

'I always believed that the job was worth it... even though only 4 beneficiaries could be reached... 4, 400, or 4 million people... it doesn't matter... their life is going to be changed forever'.

B. Showing Tolerance and Cooperation

'We are quite tolerant people, always trying to meet each other's needs'.

C. Mentally Preparing for a Mission

'I always made sure that on the plane, before we arrived in the area, people took the time to think about what they were going to go and see. Try and visualise all the horrible things that they were about to encounter, so that if and when you do encounter them they don't come as a shock to you and you're already mentally prepared for anything that you may come across. It's not just things you may see, it's the things you may hear, and things that you may smell... the whole environment you're about to go into'.

D. Mentally Distancing from Occupational Stressors

I also go and do something that requires lots of attention, like driving a truck. In (*name of country*) we had many trucks. If I felt stressed I'd go to one of the truck drivers, drive this truck the whole afternoon. I can't drive trucks... well, I'm not licensed to drive trucks so... while I was driving the truck I was so careful about not making mistakes or crashing that I'd forget about the world. I was just so concentrated on one thing the whole time'.

TABLE 3.2
Aid Workers Talking about Social Networks, Occupational Stressors and Future Plans

Examples of Impact of and on Social Networks

'But the team as well, the people that I got on with… we kept each other going I guess… when it got hard. We talked a lot and did a lot of things together, and I guess kept each other going'.

'There were some friends who were very supportive and very encouraging, very… you know… supportive'.

'I must say that the last conflict in *(name of country)* hurt me psychologically. After the *(name of country)* experience it took me a while to get back to normality in terms of relationships with my family and friends'.

When asked to identify occupational stressors, participants mainly identified the external environment (e.g. security threats, working and living in close proximity to others all the time), foreign status, excessive workloads, inconclusive pre-mission training and almost non-existent post-deployment briefing. On the other hand, even though many people reported feeling psychologically vulnerable because of their stressful jobs, many seemed to be determined to go back to the field over and over again. Quotations in Table 3.3 illustrate these key themes.

3.4.1 DISCUSSION AND CONCLUSION

As seen before, resilience describes a process of learned resourcefulness and growth, as well as the ability to function psychologically at a better level than expected given the individual's skills and previous experiences (Paton et al., 2000). Masten (2001) views it as a form of 'ordinary magic' originating from the operation of basic human adaptational systems when confronted with environmental stressors. According to the results of the investigation by Comoretto et al., LASC scores (a measure of stressful experiences), the use of mental disengagement (a coping technique), the age at which participants had left education, as well as the presence of social support networks in their lives significantly predicted changes in resilience at baseline and follow-up. These three domains of protective factors appeared to be the core active ingredients that influenced people's reactions to stressful experiences. Because protective factors are commonplace features of human nature (Fredrickson, 1998), this finding aligns with Masten's portrayal of resilience as a sort of everyday magic.

In this study, people reporting higher scores on the LASC (measuring trauma) were more likely to develop negative changes in resilience than those reporting lower LASC scores. Moreover, they were more likely to resort to the use of mental disengagement techniques to respond to stressors. In a consistent way the qualitative results showed that positive coping strategies (and NOT mental disengagement), for instance being tolerant and being able to mentally prepare for the worst, were more effective to deal with environmental stressors because they lowered perceived stress and enhanced positive changes in resilience. Findings also showed mental disengagement to be associated to negative changes in resilience. According to Pearlin (1991),

TABLE 3.3
Aid Workers Talking about Occupational Stressors

Examples of Occupational Stressors

A. Security Threats

'There was a lot of insecurity out there. It was quite unsafe. That was probably one of the hardest things for me to cope with… not feeling safe was really difficult'.

B. Working and Living in Close Proximity to Others All the Time

'We were working together all the time, we were living together and spend time together 'cause there's not really anybody else around. That's pretty intense for somebody not in a relationship, and someone you wouldn't necessarily choose as your friend at home'.

C. Foreign Status

'People know you are a foreigner… in the Arab neighbourhood they immediately think you are Jewish. From a somatic point of view I could easily be a local Jewish, or an American Jewish tourist, don't you think? Young people used to throw stones at me; they shot me using plastic rifles. They know you've just arrived in town; even though you think you look ordinary, they know you are not from here. All these factors made my life more stressful'.

D. Heavy Workloads

'The first organisation in (*name of country*) never really enforced vacation time. It was work work work, 24 hours per day literally. I had one and a half breaks during 11 months of work. Basically, in those 28–29 weeks I worked for the agency I had about 3 weeks off.

Two weeks at one time and then 1 week another time. Although I know we are resilient people and we take our jobs very seriously, it's just not healthy'.

E. Pre-mission Training and Post-deployment Briefing

'Everything depends on the organisation and on its activities. It depends on the context and on the project which constitutes the framework for the individual to work within. I believe it's fundamental to conduct good selection processes, which is the weak point of every agency. Nobody is immune. I believe selection is susceptible to mistakes. It's impossible to totally understand and judge at the same time'

'None of the organisations I've worked with have given me debriefing; none of them have talked about stress, talked about stress-related problems, so… As far as I'm concerned, none of the organisations had any structured approach to the thing'.

avoiding active confrontation with a stressor to reduce any emotional tension is more likely to result in increased stress and less resilience. Mentally disengaging from a problem often prevents active coping (Billings and Moos, 1984) and is non-adaptive over the long term (Carver et al., 1989). During interviews, participants reported mentally distancing from occupational problems as a common form of coping, too.

In consistence with previous investigations (Masten et al., 1990; Cardozo and Salama, 2002), those people who could count on the availability of social networks in this study were more likely to demonstrate positive changes in resilience. In addition, interviewed aid workers who reported being able to rely on families and friends as sources of support were more likely to positively respond to the various stressors encountered during a mission. Participants leaving school at a later age were also able to rely on a wider social support network that allowed for positive changes in resilience. Moreover, people who had spent more time in education were less likely to mentally disengage from the environmental stressors characterising missions, and therefore more likely to positively adapt to them. Finally, those who had left school at a later age were also less likely to report higher LASC scores (indicative of mental distress) and more likely to experience positive changes in resilience.

Under highly stressful circumstances such as the ones characterising humanitarian missions, processes may occur that result in long-term changes in the individual – either positive, through the development of resilience resources, or negative, through the development of vulnerabilities. Findings of this study imply that the majority of surveyed participants could overcome stressful situations naturally and over time through their intelligence and ability to rely on social relations. According to Carver (1998) benefits gained from thriving in the aftermath of trauma can be applied to new experiences and future events, leading to more effective functioning. People who survive difficult moments may learn new skills and knowledge and gain confidence and mastery in their abilities to cope with future stressful events (Aldwin et al., 1996; Park, 1998).

Aid workers are a very diverse group, ranging from school leavers to retired people, who may work alone, with a partner or as part of a team, in a conflict region or a peaceful area, for weeks or for decades. Some are involved with relief work, while others participate in development projects only. Some find the experience traumatic, while others enjoy it. Their personalities lead them to boost their self-esteem and create a sense of identity through positive feelings provoked by helping others. Nonetheless, despite their best intentions, these individuals often end up hurt by the situation in which they are thrust. Therefore, relief and development workers should be encouraged to develop an emotional style capable of managing danger by estimating potential risks and by coping with the (sometimes ungrateful) behaviours of the people who are being helped, as well as with those of fellow relief workers.

These people must be taught to take care of their own health or else they will not be able to take care of others. It might also be beneficial for both aid workers and those in contact with them to be aware that return to normality following missions can be a lengthy process. Some aid staff members sometimes return from one assignment and are sent to another within a few weeks. This may not give them sufficient time to overcome stress symptoms and may lead to problems with cumulative stress and decreased levels of resilience. Integrating the humanitarian experience into one's identity is additionally challenged by the complexity of providing a coherent meaning to one's actions during and after an assignment and also by the incongruence of one's emotions or their suppression. Furthermore, physical and moral disconnection from others in the host or home country is likely to reinforce the feeling of not belonging, putting a strain on one's social identity.

Through the development and testing of a model to account for changes in resilience, Comoretto et al. offered an important contribution to existing research in the field. The theoretical framework was developed by taking into account three types of protective markers and by assessing how these interrelated among each other. Furthermore, the relationship between dispositional, environmental and cognitive factors was considered to mediate perceived levels of stress. Resilience was therefore viewed as an evolving process reflecting the noble side of human experience rather than as a static personality trait. It was described as a transformation likely to bring out specific forces of the individual that would otherwise have remained hidden.

Likewise, the consideration of protective markers represented a move away from the study of individual personality traits towards the investigation of change processes associated with healthy personality developments. This trend towards the clarification of processes rather than fixed individual characteristics has been advocated by other authors (O'Leary, 1998; Luthar et al., 2000; Violanti, 2000) who have observed how exposure to stress may sometimes end up with positive individual growth and development rather than psychopathology. This focus on resilience despite adversity is still recent, but already quite widespread, and can count among its followers, for instance, many research groups identified under the name of *The International Resilience Project* (IRP) (http://resilienceresearch.org/research/projects/international-resilience), an organisation aimed at developing a better, more culturally sensitive understanding of how vulnerable individuals around the world effectively cope with life's adversities.

In conclusion, although no magic formula allowing for the maximisation of resilience was unquestionably found to exist, the findings of this study added to the emergent body of evidence depicting people as being able to effortlessly utilise a range of adjustment strategies to overcome stress when confronted with varying degrees of trauma and adversity. In coherence with Murphy and Moriarty's (1976) statement – 'as you encounter one stressful experience it strengthens you, like a vaccine, for a future crisis' (p. 263) – participants in this study were able to bounce back, recover, and then subsequently move forward towards new professional experiences, ready to face whatever life had to offer them.

REFERENCES

Aguerre, C. 2002. Which are the psychological factors that guarantee a satisfactory old age? *Psychology Practice*, 1, 15–27.

Aldwin, C. M., Sutton, K. J. and Lachman, M. 1996. The development of coping resources in adulthood. *Journal of Personality,* 64, 837–871.

Anthony, E. J. 1982. *L'enfant Vulnerable*. Paris: PUF.

Austin, C. N. and Beyer, J. 1984. Missionary repatriation: An introduction to the literature. *International Bulletin of Missionary Research*, 4, 68–70.

Bandura, A. 1977. Self-efficacy: Towards a unifying theory of behavioural change. *Psychological Review*, 84, 191–215.

Bandura, A. 1989. Human agency in social cognitive theory. *American Psychologist*, 44, 1175–1184.

Beigbeder, Y. 1991. *The Role and Status of International Humanitarian Volunteers and Organisations*. Dordrecht: Martinus Nijhoff Publisher.

Billings, A. G. and Moos, R. H. 1984. Coping, stress, and social resources among adults with unipolar depression. *Journal of Personality and Social Psychology*, 46, 877–891.

Block, J. and Kremen, A. M. 1996. IQ and ego-resiliency: Conceptual and empirical connections and separateness. *Journal of Personality and Social Psychology*, 70, 349–361.

Caplan, G. 1979. *Support Systems and Community Mental Health*. New York: Behavioural Publications.

Cardozo, B., Holtz, T. H., Kaiser, R., Gitwat, C. A., Ghitis, F., Toomey, E. and Salama, P. 2005. The mental health of expatriate and Kosovar Albanian humanitarian aid workers. *Disasters*, 29, 152–170.

Cardozo, B. L. and Salama, P. 2002. Mental health of humanitarian aid workers in complex emergencies. In: Y. Danieli (Ed.), *Sharing the Front Line and the Back Hills: Peacekeepers, Humanitarian Aid Workers and the Media in the Midst of Crisis.* Amityville, NY: Baywood.

Carr, K. 1994. Trauma and post-traumatic stress disorder among missionaries. *Evangelical Missions Quarterly*, 30, 246–253.

Carver, C. S. 1998. Resilience and thriving: Issues, models, and linkages. *Journal of Social Issues*, 54, 245–266.

Carver, C. S., Scheier, M. F. and Weintraub, J. K. 1989. Assessing coping strategies: A theoretically based approach. *Journal of Personality and Social Psychology*, 56, 267–283.

Comoretto, A., Crichton, N. J. and Albery, P. I. 2011. *Resilience in Humanitarian Aid Workers: Understanding Processes of Development.* Saarbrücken: Lambert Academic Publishing.

Conger, R. D., Rueter, M. A. and Elder, G. H. Jr. 1999. Couple resilience to economic pressure. *Journal of Personality and Social Psychology*, 76, 54–71.

Corneil, W., Beaton, R. and Murphy, S. 1999. Exposure to traumatic incidents and prevalence of post-traumatic stress symptomatology in urban firefighters in two countries. *Journal of Occupational Health Psychology*, 4, 131–141.

Cyrulnik, B. 1999. *Un merveilleux malheur.* Paris: Odile Jacob.

de Tychey, C. 2001. Surmonter l'adversité: Les fondaments dynamiques de la resilience. *Cahiers de Psychologie Clinique*, 1, 29–40.

de Waal, A. 1988. The sanity factor: Expatriate behaviour on African relief programmes. *Network Paper 2b.* Refugee Participation Network.

Donovan, K. 1992. *The Pastoral Care of Missionaries.* Lilydale: Commodore Press.

Durham, T. W., McCammon, S. L. and Jackson Allison, E. 1985. The psychological impact of disaster on rescue personnel. *Annals of Emergency Medicine*, 14, 664–668.

Egeland, B., Carlson, E. and Sroufe, L. A. 1993. Resilience as process. *Development and Psychopathology*, 5, 517–528.

Eriksson, C. B., Vande Kemp, H., Gorsuch, R., Hoke, S. and Foy, D. W. 2001. Trauma exposure and PTSD symptoms in international relief and development personnel. *Journal of Traumatic Stress*, 14, 205–212.

Fergusson, D. M. and Lynskey, M. T. 1996. Adolescent resiliency to family adversity. *Journal of Child Psychology and Psychiatry*, 37, 281–292.

Fonagy, P., Steele, M., Steele, H., Higgitt, A. and Target, M. 1994. The Emanuel Miller Memorial Lecture 1992: The theory and practice of resilience. *Journal of Child Psychology and Psychiatry*, 35, 231–257.

Fox, R. C. 1995. Medical humanitarianism and human rights: Reflections on doctors without borders and doctors of the world. *Social Science and Medicine*, 41, 1607–1616.

Fredrickson, B. L. 1998. What good are positive emotions? *Review of General Psychology*, 2, 300–319.

Gibbs, M. S., Drummond, J. and Lachenmeyer, J. R. 1993. Effects of disasters on emergency workers: A review with implications for training and post-disaster interventions. *Journal of Social Behaviour and Personality*, 8, 189–212.

Gore, S. and Eckenrode, J. 1994. Context and process in research on risk and resilience. In: R. J. Haggerty, L. R. Sherrod, N. Garmezy and M. Rutter (Eds.), *Stress, Risk, and Resilience in Children and Adolescents.* New York: Cambridge University Press.

Herrenkohl, E. C., Herrenkohl, R. C. and Egolf, B. 1994. Resilient early school-age children from maltreating homes: Outcomes in late adolescence. *American Journal of Orthopsychiatry*, 64, 301–309.

Horowitz, M. J. 1993. Stress response syndromes. In: J. P. Wilson and B. Raphael (Eds.), *International Handbook of Traumatic Stress Syndrome.* New York: Plenum Press.

Jones, E. S. and Jones, M. E. 1994. Strangers and exiles: Caring for the missionary. *Career and Counsellor*, 4, 32–37.

Kandel, E., Mednick, S. et al. 1988. IQ as a protective factor for subjects at high risk for antisocial behavior. *Journal of Consulting and Clinical Psychology*, 56, 224–226.

Kaspersen, M., Matthiesen, S. B. and Gotestam, K. G. 2003. Social network as a moderator in the relation between trauma exposure and trauma reaction: A survey among UN soldiers and relief workers. *Scandinavian Journal of Psychology*, 44, 415–423.

Kumpfer, K. L. 1999. Factors and processes contributing to resilience: The resilience framework. In: M. D. Glantz and J. L. Johnson (Eds.), *Resilience and Development: Positive Life Adaptations*. New York: Kluwer Academic/Plenum.

Lane, P. S. 1994. Critical incident stress debriefing for health care workers. *Omega: Journal of Death and Dying*, 28, 301–315.

Lazarus, R. S. 1999. *Stress and Emotion: A New Synthesis*. New York: Springer.

Luthar, S. 1991. Vulnerability and resilience: A study of high risk adolescents. *Child Development*, 62, 600–616.

Luthar, S. 2003. *Resilience and Vulnerability: Adaptation in the Context of Childhood Adversities*. New York: Cambridge University Press.

Luthar, S., Cicchetti, D. and Becker, B. 2000. The construct of resilience: A critical evaluation and guidelines for future work. *Child Development*, 71, 543–562.

Macnair, R. 1995. *Room for Improvement: The Management and Support of Relief and Development Workers*. London: Overseas Development Institute.

Masten, A. 1994. Resilience in individual development: Successful adaptation despite risk and adversity. In: M. C. Wang and E. W. Gordon (Eds.), *Risk and Resilience in Inner City America: Challenges and Prospects*. Hillsdale, NJ: Erlbaum.

Masten, A. 2001. Ordinary magic: Resilience processes in development. *American Psychologist*, 56, 227–238.

Masten, A., Best, K. and Garmezy, N. 1990. Resilience and development: Contributions from the study of children who overcome adversity. *Development and Psychopathology*, 2, 425–444.

Masten, A. S., Garmezy, N., Tellegen, A., Pellegrini, D. S., Larkin, K. and Larsen, A. 1988. Competence and stress in school children: The moderating effects of individual and family qualities. *Journal of Child Psychology and Psychiatry*, 29, 745–764.

Masten, A. S., Hubbard, J. J. and Gest, S. D. 1999. Competence in the context of adversity: Pathways to resilience and maladaptation from childhood to late adolescence. *Development and Psychopathology*, 11, 143–169.

McLean, S., Wade, T. D. and Encel, J. S. 2003. The contribution of therapist beliefs to psychological distress in therapists: An investigation of vicarious traumatisation, burnout and symptoms of avoidance and intrusion. *Behavioural and Cognitive Psychotherapy*, 31, 417–428.

Mitchell, J. T. and Dyregrov, A. 1993. Traumatic stress in disaster workers and emergency personnel: Prevention and intervention. In: J. P. Wilson and B. Raphael (Eds.), *International Handbook of Traumatic Stress Syndromes*. New York: Plenum Press.

Murphy, L. B. and Moriarty, A. E. 1976. *Vulnerability, Coping, and Growth from Infancy to Adolescence*. New Haven: Yale University Press.

O'Leary, V. E. 1998. Strength in the face of adversity: Individual and social thriving. *Journal of Social Issues*, 54, 425–445.

Park, C. L. 1998. Stress-related growth and thriving through coping: The roles of personality and cognitive processes. *Journal of Social Issues*, 54, 267–277.

Paton, D. 1992. International disasters: Issues in the management and preparation of relief workers. *Disaster Management*, 4, 183–190.

Paton, D., Smith, L. and Violanti, J. 2000. Disaster response: Risk, vulnerability and resilience. *Disaster Prevention and Management*, 9, 173–179.

Pearlin, L. I. 1991. The study of coping: An overview of problems and directions. In: J. Eckenrode (Ed.), *The Social Context of Coping*. New York: Plenum Press.

Pines, A. and Maslach, C. 1978. Characteristics of staff burnout in mental health settings. *Hospital and Community Psychiatry*, 29, 233–237.

Price, G. 1913. Discussion on the causes of invaliding from the tropics. *British Medical Journal*, 2, 1290–1296.

Procidano, M. E. and Heller, K. 1983. Measures of perceived social support from friends and from family: Three validation studies. *American Journal of Community Psychology*, 11, 1–24.

Raphael, B. 1984. Psychiatric consultancy in major disasters in Australia and New Zealand. *Journal of Psychiatry*, 18, 303–306.

Raphael, B., Singh, B. and Bradbury, L. 1986. Disaster: The helper's perspective. In: R. H. Moos (Ed.), *Coping with Life Crises*. New York: Plenum.

Richardson, J. 1992. Psychopathology in missionary personnel. In: K. O'Donnell and M. L. O'Donnell (Eds.), *Missionary Care*. California: William Carey Library.

Rotter, J. 1966. Generalised expectancies for internal versus external control of reinforcement. *Psychological Monographs*, 80, 3–28.

Rutter, M. 1989. Pathways from childhood to adult life. *Journal of Child Psychology and Psychiatry*, 30, 23–51.

Sarason, I. G., Sarason, B. R. and Pierce, G. R. 1995. Stress and social support. In: S. E. Hofball and M. W. De Vries (Eds.), *Extreme Stress and Communities: Impact and Intervention*. Dordrecht: Kluwer.

Scarr, S. and McCartney, K. 1983. How people make their own environments: A theory of genotype 'RA' environmental effects. *Child Development*, 54, 424–435.

Schouten, E. J. and Borgdorff, M. W. 1995. Increased mortality among Dutch development workers. *British Medical Journal*, 311, 1343–1344.

Seifer, R., Sameroff, A. J. and Baldwin, C. P. 1992. Child and family factors that ameliorate risk between 4 and 13 years of age. *Journal of the American Academy of Child and Adolescent Psychiatry*, 31, 893–903.

Seligman, M. E. P. 1991. *Learned Optimism*. New York: Knopf.

Seyle, H. 1985. History and present status of the stress concept. In: A. Monat and R. S. Lazarus (Eds.), *Stress and Coping: An Anthology*. New York: Columbia University Press.

Slim, H. 1995. The continuing metamorphosis of the humanitarian practitioner: Some new colors for an endangered chameleon. *Disasters*, 19, 110–126.

Stearns, S. D. 1993. Psychological distress and relief work: Who helps the helpers? *RPN*, 15, 3–8.

Stewart, M. and Hodgkinson, P. 1994. Post-traumatic stress reactions in the professionals. In: R. Watts and D. J. De L'Horne (Eds.), *Coping with Trauma: The Victim and the Helper*. Brisbane: Australian Academic Press.

Straker, G. 1993. Exploring the effects of interacting with survivors of trauma. *Journal of Social Development in Africa*, 8, 33–47.

Summerfield, D. 1990. *The Psychosocial Effects of Conflicts in the Third World*. London: Oxfam.

Violanti, J. M. 2000. Scripting trauma: The impact of pathogenic intervention. In: J. M. Violanti, D. Paton and C. Dunning (Eds.), *Posttraumatic Stress Intervention*. Springfield, IL: Charles C. Thomas.

Werner, E. 1993. Risk, resilience, and recovery: Perspectives from Kauai. *Developmental Psychopathology*, 5, 503–515.

Werner, E. and Smith, R. S. 1982. *Vulnerable but Invincible: A Longitudinal Study of Resilient Children and Youth*. New York: McGraw Hill.

Werner, E. and Smith R. 1992. *Overcoming the Odds: High Risk Children from Birth to Adulthood*. Ithaca, NY: Cornell University Press.

White, J. L., Moffitt, T. E. and Silva, P. A. 1989. A prospective replication of the protective effects of IQ in subjects at high risk for juvenile delinquency. *Journal of Consulting and Clinical Psychology*, 57, 719–724.

World Health Organisation (WHO) 1992. *Psychosocial Consequences of Disasters: Prevention and Management*. Geneva: World Health Organisation.

4 Entrepreneurial Resilience

Role of Policy Entrepreneurship in the Political Perspective of Crisis Management

Lee Miles and Evangelia Petridou

CONTENTS

4.1 INTRODUCTION

Over the last few decades, considerable attention has been paid by scholars in political science and other disciplines to the frequency and the impact of long-term change in the international environment. Increasing in frequency and scope, multi-scalar demands have been placed at the doors of policy makers at the regional, national and international levels to develop and revise strategies in coping with key events and critical junctures affecting global and local politics. Such attention has been enhanced with recurrent debates – such as the one on climate change – that are accompanied by a further focusing of the minds of scholars and practitioners on the regular instances of unexpected consequences of change brought about by means of man-made or natural disasters.

Concomitantly, there has been an explosion of scholarship on crisis management and emergency planning rendering them important sub-fields of study for researchers across the social sciences and beyond. As Wilkenfield et al. (2005, p. 1) argue, '[r]ecognizing the primacy of crises as defining moments in international relations, scholars and policy makers alike are increasingly concerned with identifying mechanisms for crisis prevention, management and resolution'. Furthermore, a considerable segment of recent literature in a variety of disciplines (political science, sociology, disaster studies, geography, planning, to name a few) has been devoted to exploring resilience, a term that seems to have usurped sustainability both in conceptual terms for scholars and as the desired goal for practitioners. Of course, resilience, like sustainability and even crisis (see Smith 2006), suffers from a high degree of conceptual ambiguity since 'defining *resilience* is complicated by the fact that various disciplines employ the terms in slightly different ways... often with little understanding of the primary actors and institutions involved in its application' (Egli 2014, p. 4). Nevertheless, there have been some attempts at addressing this gap; such work on resilience has, for example, focused on conceptualising why and how agents, institutions and structures can bounce back and recover quickly from unexpected shocks and more recently has sought to understand how resilience can be seen as an underlying concept enabling policy makers to handle change and turbulence as part of business-as-usual, life and well-being.

An implicit understanding within the resilience literature is that there must be space to adapt quickly to all aspects of turbulence, crisis and shock. In the context of this chapter then, it is argued that works on resilience – whether consciously or unconsciously – have assumed that there must be space for change to take place. In other words, resilience is another way to think about change. As Walker and Salt (2006, pp. 9–10) observe, 'at the heart of resilience thinking is a very simple notion – things change – and to ignore or resist this change is to increase our vulnerability and forego emergency opportunities. In doing so, we limit our options'. In the political science literature, a special kind of political actor has been identified as an agent of change: the policy entrepreneur, identified as the 'political [actor] who seek[s] policy changes that shift the *status quo* in given areas of public policy' (Mintrom 2015, p. 104).

In some respects, the fact that crises open up a window – indeed a space in which policy reforms can take place – is not a new claim, but according to 't Hart and Boin (2001) one of the most under-researched elements of crisis management. Equally, it is essential that there be greater understanding of the role of these actors and entities and the reasons behind variations in response, especially, since, as Walker and Salt suggest, 'humans are usually good at noticing and responding to change. Unfortunately, we are not so good at responding to things that change' (Walker and Salt 2006, p. 10).

In particular and drawing from 't Hart and Boin (2001) (also Boin et al. 2008), we argue that there are notable linkages among the political aspect of crisis management, resilience and policy entrepreneurship. More specifically:

1. 't Hart and Boin outline the two aspects of crisis termination: the managerial one, 'which is all about the functional adaptation of communities, administrative agencies and political decision-making processes to the

extreme conditions of crisis' ('t Hart and Boin 2001, p. 29) and the political one, a kind of stage in which political actors, incumbents and other stakeholders alike play out the drama of policy making.

2. Boin and McConnell (2007) suggest that crisis management increasingly incorporates principles of resilience and they go on to introduce the concept of 'societal resilience' as a theoretical construct with which to understand the 'bouncing back' in the managerial perspective of crisis management. This is the resilience of citizens and practitioners who ensure that the system (specifically critical infrastructures) returns to a functioning state.

3. The political aspect of crisis management consists of the political maelstrom during, and specifically after, a crisis, which breaks the crisis open and exposes it for what it is – an emerging entrepreneurial space for decision making. It should be noted that this decision-making arena not only has to allow for 'bouncing back', but also to enable the emergence of entrepreneurial solutions in order to respond, often quickly, to unexpected consequences of change as well as pursue future change. This assumption builds upon what Wildavsky (1988, p. 2) argued – namely that there is a need 'to redress this imbalance [that the search for safety has been identified as keeping things from happening] by emphasising the increases in safety due to entrepreneurial activity'. In the context of this chapter then, it is argued that effective crisis management has to incorporate the principle of what we label as 'entrepreneurial resilience'.

4. This arena of 'entrepreneurial resilience' can be interpreted as an opportunity structure for policy entrepreneurs and policy making, characterised by a sense of bouncing *forward* (Davoudi 2012). Entrepreneurial actors can act as change agents proposing and implementing innovative solutions to remedy and alleviate the consequences of (ongoing) crisis events. They can also contribute with solutions and formulate policies in order to reduce the possibility of the advent of certain crises as well as lower the chance of failure when handling future crises events.

In sum, the synergetic ideas of evolutionary resilience and the political and managerial aspects of crisis management, both separately and collectively, assume that a space for entrepreneurship and change must exist as part of any successful application of evolutionary resilience and crisis management.

The purpose of this chapter is thus twofold. First, we outline how resilience and crisis management can be understood as encapsulating an implicit acceptance of change and thereby discuss how entrepreneurship and change can be understood – alongside and within – notions of resilience. In particular we introduce the notion of 'entrepreneurial resilience' as central to the political perspective of crisis management. Second, the chapter introduces and discusses how political entrepreneurship – and more specifically the role of entrepreneurs – can provide extra added value in understanding how these resilient (policy) entrepreneurs act as change agents in the arena for change.

4.2 RESILIENCE IN CONTEXT

Prindle (2012) warns us of the pitfalls inherent in borrowing concepts from other disciplines. Descriptive models originating in the natural sciences can serve as powerful heuristics; however, they do not adequately account for human agency. As powerful as such models are in providing mechanical analogies, they often do not provide a causal mechanism linking human agency ('the result of conscious beings making choices') to output (Prindle 2012, p. 37). In similar fashion, resilience has suffered from conceptual fuzziness due to its origins in ecology, subsequent transdisciplinary travel, ensuing trendiness and application to a variety of contexts in social sciences (Markusen 1999). These contexts, in the form of modifiers (from ecological to community and organisational) slightly tweak the concept to fit respective idiosyncrasies (for an example of a list of terms, see McAslan 2010). What is more, resilience is also an approach and a science (Walker and Cooper 2011) but also in practical terms, an embedded strategy in emergency preparedness guidelines.

The term was first introduced in ecology and the environment by Holling in his seminal 1973 paper. Holling, a conservation ecologist, distinguishes between engineering resilience and ecological resilience. The former is considered to be the ability of a system to return to an earlier state of equilibrium after a shock. The measure of engineering resilience is the time (t) it takes this system to return to the previous state of equilibrium and the focus, indeed, is on equilibrium. The latter – ecological equilibrium – is the degree of shock a system can absorb before it breaks down. Conversely, the emphasis here is not on one single equilibrium state as a desired or achievable state.

The ubiquity of these shocks and the multiplicity and non-linearity of their triggers has led to theorists constructing a more holistic concept of a system with continuous feedback loops and multiple equilibria. More specifically, Davoudi (2012, p. 302) elaborates on the perspective of evolutionary resilience, suggesting that 'faced with adversities, we hardly ever return to where we were'. Rather than focusing on a return to an elusive normalcy, resilience is seen as the ability of systems to change, adapt and transform. Indeed, as often quoted, 'the past is a foreign country' (Lowenthal 1985) and not a state to which a system can, or sometimes even would want to, return. Pendall et al. (2010) remind us that the social processes of pre-Katrina New Orleans is not a state to which its residents would desire to return (Davoudi 2012). In exploring the notion of economic resilience, Simmie and Martin (2010) reject equilibria altogether arguing that the economic development of regions is based on drivers such as knowledge, innovation and capitalism, which are by definition incompatible with stasis and equilibrium.

Despite the apparent lack of convergence in the literature as to the ontology of resilience, common characteristics emerge. In a review of the concept with a view to [u]nderstanding its origins, meaning and utility, McAslan (2010) posits that disparate definitions notwithstanding, common characteristics include the ability to bounce back from an adverse extraordinary event (see e.g. Holling 1973; Wildavsky 1988); preparedness to handle extraordinary events potentially causing major disturbances (see e.g. Bhamra et al. 2011); an ability to adapt to an often difficult environment (see e.g. Davoudi 2012); a tenacity and will to survive (see e.g. Norris et al., 2008) and a willingness of community members to mobilise for a common cause (Linell 2014).

Indeed, the definition which captures resilience at the most reduced and yet succinct level is by Walker and Cooper (2011, p. 154): '[w]hat is resilience, after all, if not the acceptance of disequilibrium itself as a principle of organization?' In other words, resilience is another way to conceptualise change. This is the departure point of our argumentation linking resilience and policy entrepreneurship in the field of crisis management.

4.3   ON CRISIS MANAGEMENT: A SPACE FOR RESILIENT ENTREPRENEURS

Shocks that can potentially destabilise systems can take many forms (a flood, a war, an invasive species, a terrorist attack, a school shooting, sub-prime mortgage implosion) depending on the definition of the system, but we tend to view these shocks as crises. In terms of policymaking, Boin et al. (2005, p. 2) define a crisis as 'a serious threat to the basic structures or the fundamental values and norms of a system, which under time pressure and highly uncertain circumstances necessitates making vital decisions'. Despite the intuitive thinking that crises have a direct large exogenous trigger, Boin et al. (2005) posit that the causes of crises are incremental and they lie within the system itself. This incremental nature of the mechanisms behind the emergence of crises contributes to their ubiquity and unpredictability. In other words, there is no way we can predict the next disaster; the best we can do is construct a system able to bounce back after a crisis: a resilient system. In mainstream resilience literature, it is given that 'there must be a general awareness that a catastrophe may strike, paralysing normal governmental functions and CIs.* (Boin and McConnell 2007, p. 54). Resilience as an operational strategy in emergency preparedness, crisis response and national security has the underlying assumption that safety is unattainable. The state is unable to keep citizens safe by preventing crises and identifies resilience as the main component of a culture of preparedness (Walker and Cooper 2011). Though this has been largely unproblematised in a considerable portion of the resilience literature, the accentuated relationship of the concept of resilience with late liberal ideas has been addressed in critical texts (see e.g. Chandler 2013, 2014; Zebrowski 2013), especially the political implications the apotheosis of resilience has while the state promises an unequivocal lack of safety to its subjects (Evans and Reid 2014).

In order to further investigate crisis as an abstract concept, it would be fruitful to turn to linguistics and the literal meaning of the word. Originating from the Greek 'κρίσις', the word originally meant 'separating, power of distinguishing', 'a decision, judgment'; in legal terms 'a trial' or 'the result of a trial', as well as another kind of trial, a 'trial of skill', for example in archery (Liddell and Scott 1889). It is no wonder then that we associate crises nowadays with trials challenging the constancy of all the structural factors that allow us to live our lives. What is also an integral part in 'crisis' is the element of decision, that is, the ability to distinguish between different

---

* CIs: Critical Infrastructures.

elements and make a judgment. This is reflected in the emphasis we put on the role of leadership in times of crises.

Furthermore, Smith (2006, p. 6) argues that notions of crisis incorporate generic factors such as issues of place, time, emergence and scale that ensure that complex, non-linear problems are faced by crisis managers; and in many ways, crisis events illustrate and spotlight inherent vulnerabilities that can exist within organisations (see also Smith 2005) and need to be 'fixed'. Crises then, on a political and institutional level and in terms of policy making, present opportunity junctures. They are arenas for decisions made with the goal of exiting the crisis that is to say, aiming at change. However, crises are not 'one-and-done' affairs, neatly delineated spatially and temporally ('t Hart and Boin 2001). Dror (1993, p. 13) explicitly recognises the dual nature of crises as opportunities (as well as burdens) 'because of the softening of institutional rigidities which, consequently, open up new vistas for political or social feasibility' and claims that 'crisis coping must be combined with opportunity handling' and indeed the upgrade of the steering capacity of governance structures to handle crises–opportunities is crucial to statecraft and enterprise building. In other words, in current thinking resilience is the way systems deal with a continuous state of flux, within which reside incremental triggers for disturbances – crises. These crises create windows of opportunity for the making of innovative reform policies. Indeed, '[i]n a resilient social–ecological system, disturbance has the potential to create opportunity for doing new things, for innovation and for development' (Folke 2006, p. 253). Moreover, official resilience reports in recent years have increasingly recognised the critical role played by a robust coalition of actors and leaders (in operationally delivering resilience during peaks in crises events) that 'have at its core a strong leadership and governance structure, and people with adequate time, skills and dedication' (National Academies 2012, p. 6).

If this is the case, it is all the more surprising that resilience studies have not always connected with other discourses in the social sciences that have also attracted the attention of scholars over the past decades, such as studies of entrepreneurship. Having said this, Hogan and Feeney (2011) make the connection between political/policy entrepreneurs and crisis management arguing that the notion that exogenous shocks are automatically the causes of policy change is too simplistic. They go on to say that the agency of entrepreneurial teams introduce and transform new ideas into policy in the wake of a crisis. We take this one step further and argue that a better understanding of political agency is the key to understanding policy change (also Folke et al. 2005) and the concept of resilience in the political aspect of crisis management.

In the urban governance literature, Williams et al. (2013) note the synergies of market entrepreneurship and the economic resilience of city regions. In attempting to answer the question of why some regions are more (economically) resilient than others, the authors point out that the diversity and flexibility of entrepreneurs – critical to competitiveness and growth – are also sources of resilience to external shocks. They conclude that (market) entrepreneurship is a dynamic driver of resilience. What is more, an implicit (though not direct) connection between market entrepreneurship and resilience is made by Burnard and Bhamra (2011) in the context of small and medium-size enterprises (SMEs). As Narbutaité-Aflaki et al. (2015) have observed elsewhere, the study of entrepreneurship – under the broad banner of political/policy entrepreneurship – has increased in both scope and size in recent decades, traveling through disciplines including economics, organisational studies, administrative sciences, policy studies and political science. These works on political/policy entrepreneurship have – at the risk of some generalisation – sought to develop more nuanced understandings of the role of agents, institutions and structures in initiating and implementing change. At first glance, and as explored further here, there would seem to be an apparent nexus and potential for substantial interaction between resilience research and policy entrepreneurship research in the larger context of crisis management that is clearly worth further investigation. In short, detailed insight from the world of political entrepreneurship can provide us with more sophisticated tools for understanding how entrepreneurial activities (i.e. of resilient entrepreneurs) operate within resilient decision-making processes of crisis management and emergency planning. In effect, such insights would shed light on practices of resilient entrepreneurs facilitating 'bouncing forward' in order to enable entrepreneurial resilient systems to 'bounce back'.

4.4 CHANGE AND ENTREPRENEURSHIP IN RESILIENT POLICY-MAKING PROCESSES

4.4.1 POLICY ENTREPRENEURSHIP

Entrepreneurship, widely recognised as the motor of capitalism, has been investigated at length in disciplines such as economics, business administration and regional development studies, to name but a few. What is more, the concept of political entrepreneurship has been gaining ever increasing momentum during the recent years with a body of literature increasing in a variety of disciplines, particularly in political science (see e.g. Ostrom 1965; Schneider et al. 1995; Mintrom 1997, 2000; Sheingate 2003; Mintrom and Norman 2009; Martin and Thomas 2013). Simply put, 'policy entrepreneurs are political actors who seek policy changes that shift the *status quo* in given areas of public policy' (Mintrom 2015, p. 103). Sheingate takes this one step further by noting that 'entrepreneurs are individuals whose creative acts have transformative effects on politics, policies or institutions' (Sheingate 2003, p. 185). In their quest for change, entrepreneurs in the public sphere must be embedded in the system they seek to change, they must be alert in order to discover opportunities in the form of unfulfilled needs, be willing to take the risk involved in this exploration of opportunities and also must be able to amass coalitions in order to push for change (Schneider et al. 1995; Petridou et al. 2015).

Even though early studies focused on the policy entrepreneur as the exceptional, larger than life individual (see e.g. Lewis 1980; Loomis 1988), later studies show that entrepreneurship is not a quality intrinsic to individuals – that is to say it is not a matter of one being entrepreneurial in the same way as one having blue or brown eyes. Rather, policy entrepreneurship comprises the set of behaviours mentioned above and is highly contextualised (see Narbutaité-Aflaki et al. 2015). Any one person can act as a policy entrepreneur in a particular context and policy sector. What is more, and even though we are discussing entrepreneurship in the public sphere, the actor does not necessarily have to be a politician; he can be a bureaucrat, part of the non-profit sector or a concerned citizen at any level of governance. A common feature of all these actors is that their goal is affecting change.

Being a policy entrepreneur does not ensure success in affecting change and we can learn valuable lessons from failed attempts of policy entrepreneurs (see e.g. Hays 2012; Mintrom 2015). At the same time, most research of policy entrepreneurship is qualitative, based on case studies examining change and working backwards to trace the entrepreneurial mechanisms contributing to this change. An exception to this is the U.S.-wide study by Michael Mintrom (2000) on educational policy at the state level. Partly due to this focus on the observable (change) has caused policy entrepreneurship to be investigated as the explanatory variable vis-á-vis shifts in the *status quo*. Conversely, the structural factors and their potential to hinder or enable entrepreneurship are relatively under researched. In this chapter, we argue that during the political side of crisis management, structural arrangements become flexible thus allowing more space for policy entrepreneurs to act.

Policy making at times of crises is not a new field of enquiry in political science and policy studies as crises are seen as exogenous, destabilising shocks (see Birkland, 1998 for a study on focusing events and agenda setting). The added value of exploring the political perspective of crisis management and resilience through the lens of policy entrepreneurship is the understanding of exceptional agency in extraordinary contexts.

Nevertheless, a unitary, comprehensive approach to understanding, analysing and explaining policy change is neither necessary or indeed, adequate (see also Capano and Howlett 2009; John 2012; and for recent reviews of the modelling the policy process – see Nowlin 2011; Petridou 2014). Rather, we highlight that the concept of change is integral to the policy-making process, policy entrepreneurship and resilience; on the one hand, as noted above, all policy making is change and on the other, the adaptive cycle of resilience specifically describes stages of change in the structure and functions of a system (Gunderson and Holling 2002). This change can be incremental or sudden and may or may not be innovative. However, for a system to be resilient, it has to be able to change. A shock occurring, a crisis exposes faults in the system that have to be addressed if the system is to adapt and survive. Similarly, social processes exposed by crises call for innovative decisions resulting in policies able to deal with contemporary problems. It would be thus fair to assume that the more innovative the adaptation (the change), the more robustly the system bounces back as well as forward.

4.4.2 CONCEPTUALISING ENTREPRENEURIAL RESILIENCE: WHY DOES RESILIENCE HAVE A SPACE FOR ENTREPRENEURSHIP?

As Landford et al. (2010) highlight, scholars of crisis management and resilience – especially those since Hurricane Katrina in 2005 and Deepwater Horizon in 2010 – have increasingly highlighted the role of leadership in understanding crises and successful crisis management. Put simply, lessons from investigating crisis events illustrate how gaps in disaster resilience appear, gaps that must be filled by agencies and agents and largely complement the work of government structures and policy making (Landford et al. 2010). Since these actors are not – or at least not to the same degree – constrained by administrative rules or red tape as governmental bureaucracy, they 'are often more innovative or flexible than established governmental disaster relief… a trait that serves them well in the quickly changing overall environment and the specific circumstances of post-disaster situations' (Landford et al. 2010, p. 122). From the perspective of this chapter then, successful attempts at resilience often are so because of entrepreneurial actions that fill in the gaps left within and by resilient systems and planning in order to meet the challenges of crises at the operational level. As Atkinson (2014, p. 90) observes, it is often the case that 'there have been lapses of policy, the staff that were on the front line seemed to know best practices'. Put simply, successful resilience and crisis management in practice already recognises that it needs to balance coordination and communication planning priorities with the need to enable innovation and entrepreneurship on the ground in handling crisis.

Indeed, this can be taken further since Landford et al. (2010) in their study of community resilience and Hurricane Katrina illustrate the extent to which agency now features as a key future research agenda for studies of effective crisis management. The authors outline five key lessons from their investigation of Hurricane Katrina – each of which demonstrates a rationale for why resilience (and the context of resilience) already acknowledges that there must be a key space for both entrepreneurs and entrepreneurship. These are

1. *That one of the greatest obstacles to effective resilience planning and in handling crises is complacency,* which must be continually combatted. While resilience planning goes some way, understanding the role of key actors and entrepreneurs in reducing such complacency by offering political entrepreneurial solutions and promoting operational change can be important.
2. *That resilience planning must be integrated into transitions in leadership.* The Landford et al. (2010) study of Hurricane Katrina highlighted that changes in government had left inexperienced leadership in place at the time of the hurricane. It may be important then to examine how political entrepreneurs can play a role in offsetting transitions and weaknesses in leadership by promoting policy change, being agile and adaptive.
3. *That individual leadership matters.* 'Resilience requires the ability to react and command situations that are largely unexpected and no plan really accounts for. This can only be done if individual leadership is empowered to take command and understand the role and limitations of what must be done and what cannot be done. … Strong leadership can help overcome the unexpected social and logistical hurdles that have not been accounted for in advance' (Landford et al. 2010, p. 134). There is understanding that resilience systems require specific leadership skills and forms of entrepreneurial action.
4. *Resilience policy making often involves politically arduous territory.* 'In order to implement a strong resilience policy it often takes a great deal of vision and courage' (Landford et al. 2010, p. 134). There is implicit recognition in resilience that crisis management is a fundamentally politically contested policy-making arena, which is where political entrepreneurs operate.
5. *Government and the private sector are inevitably limited in their responsiveness to disaster events.* Individuals have a significant role in

(community) resilience (Landford et al. 2010, p. 136). Resilience 'inevitably must begin with the individual… resilience players must clearly communicate the limitations that are involved in what is doable and what is not and that such capabilities are only as powerful as the individuals empowered by them'. There is recognition of the role of individual agency in the resilience context as a key variable in delivering successful crisis management.

In many ways then, we are now on the road to understanding *why* resilience studies recognise the importance of actors and agency and indeed, of emphasising an important role for entrepreneurship in resilience planning and implementation. As Aldrich (2012, p. 166) highlights, part of the key to building resilience is that 'social networks are more resilient than buildings. Buildings crumble, but human resources remain'. There is therefore a clear case for understanding why the resilience context has a space for entrepreneurship.

Furthermore, this chapter argues that there is a need to understand the entrepreneurial characteristics of resilience, and there is a need for further analysis in this regard. In particular, the concept of entrepreneurial resilience represents the acceptance that there exists entrepreneurial space where individuals and collectivities are able to undertake entrepreneurial actions that fill in the gaps and affect change in order to meet the challenges of upcoming or present crisis situations. Put simply, there are (political) places and times where entrepreneurial resilience operates in practice within resilient policy making and systems. However, with the concept of entrepreneurial resilience now established, what it perhaps further lacking is a clear understanding how such political entrepreneurship operates in practice and if, when and where it influences successful aspects of crisis management and resilience. An outline of how political entrepreneurship relates to the resilience context is provided in Table 4.1 and elaborated below.

Policy entrepreneurs by definition are political agents aiming at affecting change. The success or their efforts are not predetermined and lessons can be learnt from failures. Regardless of the result of policy entrepreneurial efforts, the existence of such efforts combats complacency, which is tied to the *status quo* and business-as-usual. The creativity and dynamism of policy entrepreneurs can act as the impetus to handle complacency and undertake change.

Often, policy entrepreneurs are in leadership positions; in fact some scholars reject the notion of a policy entrepreneur as somebody who is not a leader (Capano and Howlett 2009). In any event, the policy entrepreneur must possess leadership skills; a fundamental behavioural trait of the policy entrepreneur is the ability to organise others in coalitions (partly through persuasive arguments) and provide leadership, inspiration and vision to these coalitions. Mintrom notes the importance of establishing an organisational culture as it 'holds the promise of improving cooperation among group members, giving people a clear sense of the values and general practices of the organisation and ensuring the lines of communication remain open' (Mintrom 2000, pp. 105–106). This ability, as well as leading by example are elements that can integrate leadership and most importantly fill the leadership gap if discontinuity occurs. Conversely, it is the creativity and insightfulness of

TABLE 4.1

The Nexus of Resilience and Policy Entrepreneurship

Resilience Context (Why) (Landford et al. 2010)	Policy Entrepreneurship (How) (Mintrom 2000)
Handling complacency, undertaking change	Policy entrepreneurs as change agents; actors focused on transforming the status quo
Integrating leadership, transition in leadership	Establishing organisational culture
	Coalition formation (cooperation, clear lines of communication)
	Persuasive arguments
	Lead by example
Identifying individuals that matter	Coalition formation/groups
	Identifying opportunities
	Creativity, insightfulness
Negotiating politically arduous territory	Argue persuasively
	Familiarity with settings
	Lead by example
	Taking risks
	Mobilising resources
Setting limitations and boundaries for the doable	Argue persuasively
	Coalition formation/groups
	Reframing problems and/or solutions
	Setting the agenda

entrepreneurs and the ability to discern and take advantage of opportunities that can enable them to identify individuals that matter.

Leadership skills are also vital in negotiating arduous – in other words controversial – territory. Persuasive arguments have the ability to frame issues one way or another and set agendas. Being familiar with the political situation, that is, being an insider, has proven to be an asset for a political entrepreneur (see e.g. Böcher 2015). Political know-how in terms of mobilising resources and taking political risks is also well documented in policy entrepreneurship studies (Narbutaité-Aflaki et al. 2015).

Finally, the ability of the policy entrepreneur to control the discourse and set the agenda, partly through coalition formation and persuasive arguments can help set the limits of what is doable (see Miles 2015). In summary, we posit that it is the various aspects of political/policy (resilient) entrepreneurship that can take over in crisis management and contribute toward an overall resilient process.

4.5 ENTREPRENEURIAL RESILIENCE AND POLICY ENTREPRENEURSHIP: FUTURE APPLICATIONS AND RESEARCH AGENDAS

Clearly then, it would seem fruitful to further develop the concept of entrepreneurial resilience. There is a notable basis for such endeavour precisely because this chapter has firstly demonstrated that there are obvious points of departure within the existing literature on resilience and crisis management that stress a need for further work on the importance of actors and agents within successful resilience and crisis management policy making (also Folke et al. 2005).

Secondly, there are fruitful cross-overs in the way entrepreneurship and resilience accept as part of their operational practice an essential role for policy change, and although not exclusive, agent driven change. In many ways, entrepreneurship and resilience are handling similar challenges. Entrepreneurship can help understand the opportunities presented within resilience policy making (see Lui et al. 2011). At the very least, it is not a case of entrepreneurship *or* resilience, but rather a case of the need to understand entrepreneurship *and* resilience, especially if we are to fully harness the dynamism of bouncing forward. Above all, with the developing concept of entrepreneurial resilience, we can provide further insight into the role that entrepreneurship plays in helping deliver resilience, and likewise, the role of resilience motives and assumptions in driving the ambitions of, and shaping behaviour among, entrepreneurs during the prevention, mitigation, response and recovery phases of crisis management situations.

Thirdly, this chapter goes further in arguing that there are new, fertile grounds for examining how works on resilience and crisis management, particularly those that place emphases on agents and innovation, and the literature and tool-box (Petridou et al. 2015) on political entrepreneurship inter-relate, and in particular the way in which resilience and crisis management alongside political entrepreneurship contribute jointly to mutual understandings of the why and how of entrepreneurial resilience in practice. This approach should go some way to meeting the challenge set out by Comfort et al. (2010, p. 281), that resilience works need to further understand the precepts for leadership and understand the roles of 'a solid cohort of resilience-oriented leaders'; in essence, only after further conceptual development can it be really possible to offer comprehensive suggestions about resilience leadership since 'the issue is too important to get it wrong'.

The potential for entrepreneurial resilience and resilience-political entrepreneurship in practical terms is notable. One immediate point would be to elaborate further and map out the dynamics of entrepreneurial resilience as they operate within the specific phases of emergency and crisis management (prevention, mitigation, response, recovery) and provide a more nuanced picture of the role of agents and entrepreneurial spaces in key aspects and modes of crisis management.

At the very least, it would be highly beneficial if this became an essential part of future research agendas. Too many times, scholars and practitioners of crisis management and resilience identify key actors and agents as single points of failure, whether correctly or as a part of 'blame games' (Boin et al. 2005, p. 103) – be they technical advisors and managers, for example nuclear test and plant managers during the 1986 Chernobyl nuclear disaster, senior policy makers like President Bush during Hurricane Katrina or corporate managers as a part of the 2010 Deepwater Horizon spillages. Similarly, on other occasions key individuals or groups are identified as being central to the success of episodes in crisis management and in removing or alleviating key vulnerabilities during crisis situations and as a part of ongoing resilient policy making. Moreover, it is essential that scholars develop sophisticated understandings of why this is the case, since after all, as Boin et al. (2008, p. 13) observe, 'the line between political winners and losers is such a fine one'. By thinking more clearly about the role of political entrepreneurs in handling crisis and in making for successful crisis management during all phases of emergency planning and resilience, analysts will gain deeper as well as more nuanced insights on the holy grail of charting paths to bouncing back and bouncing forward from crisis.

REFERENCES

Aldrich, D.P. *Building Resilience: Social Capital in Post-Disaster Recovery.* Chicago: The University of Chicago Press, 2005.

Atkinson, C.L. *Toward Resilient Communities: Examining the Impacts of Local Governments in Disasters.* London: Routledge, 2014.

Bhamra, R., S. Dani and K. Burnard. Resilience: The concept, a literature review and future directions. *International Journal of Production Research* 49(18), 2011: 5375–5393. doi: 10.1080/00207543.2011.563826.

Birkland, T.A. Focusing events, mobilization, and agenda setting. *Journal of Public Policy* 18(01), 1998: 53–74.

Böcher, M. The role of policy entrepreneurs in regional government processes. In: I. Narbutaité-Aflaki, E. Petridou and L. Miles (Eds.), *Entrepreneurship in the Polis: Understanding Political Entrepreneurship.* Aldershot: Ashgate, 2015, pp. 73–86.

Boin, A. and A. McConnell. Preparing for critical infrastructure breakdowns: The limits of crisis management and the need for resilience. *Journal of Contingencies and Crisis Management* 15(1), 2007: 50–59. doi: 10.1111/j.1468-5973.2007.00504.x.

Boin, A., A. McConnell and P. 't Hart. *Governing After Crisis: The Politics of Investigation, Accountability and Learning.* Cambridge: Cambridge University Press, 2008.

Boin, A., P. 't Hart, E. Stern and B. Sundelius. *The Politics of Crisis Management: Public Leadership under Pressure.* Cambridge: Cambridge University Press, 2005.

Burnard, K. and R. Bhamra. Organisational resilience: Development of a conceptual framework for organisational responses. *International Journal of Production Research* 49(18), 2011: 5581–5599. doi: 10.1080/00207543.2011.563827.

Capano, G. and M. Howlett. Conclusion: A research agenda for policy dynamics. In: G. Capano and M. Howlett (Eds.), *European and North American Policy Change* New York: Routledge, 2009, pp. 217–231.

Chandler, D. Editorial. *Resilience* 1(1), 2013: 1–2. doi: 10.1080/21693293.2013.765739.

Chandler, D. Beyond neoliberalism: Resilience, the new art of governing complexity. *Resilience* 2(1), 2014: 47–63. doi: 10.1080/21693293.2013.878544.

Comfort, L.K., A. Boin and C.C. Demchak. Resilience revisited: An action agenda for managing extreme events. In: L. K. Comfort, A. Boin and C. C. Demchak (Eds.), *Designing Resilience: Preparing for Extreme Events.* Pittsburgh: University of Pittsburgh Press, 2010, pp. 272–284.

Davoudi, S. Resilience: A bridging concept or a dead end? *Planning Theory and Practice* 13(2), 2012: 299–333.

Dror, Y. Steering requisites for crises-opportunities: On-going challenges. *Journal of Contingencies and Crisis Management* 1(1), 1993: 13–14.

Egli, D.S. *Beyond the Storms: Strengthening Homeland Security and Disaster Management to Achieve Resilience.* Armonk, NY: M.E. Sharp, 2014.

Evans, B. and J. Reid. *Resilient Life: The Art of Living Dangerously.* Malden: Polity, 2014.

Folke, C. Resilience: The emergence of a perspective for social–ecological systems analyses. *Global Environmental Change* 16(3), 2006: 253–267. doi: http://dx.doi.org/10.1016/j.gloenvcha.2006.04.002.

Folke, C., T. Hahn, P. Olsson, and J. Norberg. Adaptive governance of social-ecological systems. *Annual Review of Environment and Resources* 30, 2005: 441–473. doi: 10.1146/annurev.energy.30.050504.144511.

Gunderson, L.H. and C.S. Holling (Eds.), *Panarchy: Understanding Transformations in Human and Natural Disasters.* New York: Island Press, 2002.

Hays, B.D. The curious case of school prayer: Political entrepreneurship and the resilience of legal institutions. *Politics and Religion* 5, 2012: 394–418.

Hogan, J. and S. Feeney. Crisis and policy change: The role of the political entrepreneur. *Risk, Hazards and Crisis in Public Policy* 3(2), 2011: 1–24.

Holling, C.S. Resilience and stability of ecological systems. *Annual Review of Ecological Systems* 4, 1973: 1–23.

John, P. *Analyzing Public Policy*, 2nd ed. New York: Routledge, 2012.

Landford, T., J. Covarrubias, B. Carriere and J. Miller. *Fostering Community Resilience: Homeland Security and Hurricane Katrina.* Aldershot: Ashgate, 2010.

Lewis, E. *Public Entrepreneurship: Toward a Theory of Bureaucratic Power.* Bloomington: Indiana University Press, 1980.

Liddell and Scott. *An Intermediate Greek Lexicon.* 7th ed. Oxford: Oxford University Press, 1889.

Linell, M. Citizen response in crisis: Individual and collective efforts to enhance community resilience. *Human Technology* 10(2), 2014: 68–94.

Loomis, B. *The New American Politician: Ambition, Entrepreneurship, and the Changing Face of Political Life.* New York: Basic Books, 1988.

Lowenthal, D. *The Past is a Foreign Country.* Cambridge: Cambridge University Press, 1985.

Lui, A., R.V. Anglin, R.M. Mizelle Jr. and A. Plyer (Eds.), *Resilience and Opportunity: Lessons from the US Gulf Coast after Katrina and Rita.* Washington: Brookings Institution Press, 2011.

Markusen, A. Fuzzy concepts, scanty evidence, policy distance: The case for rigour and policy relevance in critical regional studies. *Regional Studies* 37, 1999: 701–717.

Martin, A. and D. Thomas. Two-tiered political entrepreneurship and the congressional committee system. *Public Choice* 154(1–2), 2013: 21–37. doi: 10.1007/s11127-011-9805-z.

McAslan, A. *The Concept of Resilience: Understanding its Origins, Meaning and Utility.* [Online Strawman Paper] Accessed 29 October, 2010. http://torrensresilience.org/images/pdfs/resilience%20origins%20and%20utility.pdf.

Miles, L. Political entrepreneurs as painful choices: An examination of Swedish (post)-neutrality policy. In: I. Narbutaité-Aflaki, E. Petridou and L. Miles (Eds.), *Entrepreneurship in the Polis: Understanding Political Entrepreneurship.* Aldershot: Ashgate, 2015, pp. 133–149.

Mintrom, M. Policy entrepreneurs and the diffusion of innovation. *American Journal of Political Science* 41(3), 1997: 738–770.

Mintrom, M. *Policy Entrepreneurs and School Choice.* Washington: Georgetown University Press, 2000.

Mintrom, M. Policy entrepreneurs and morality politics: Learning from failure and success. In: I. Narbutaité-Aflaki, E. Petridou and L. Miles (Eds.), *Entrepreneurship in the Polis: Understanding Political Entrepreneurship.* Aldershot: Ashgate, 2015, pp. 103–117.

Mintrom, M. and P. Norman. Policy entrepreneurship and policy change. *The Policy Studies Journal* 37, 2009: 649–667.

Narbutaité-Aflaki, I., E. Petridou and L. Miles (Eds.), *Entrepreneurship in the Polis: Understanding Political Entrepreneurship.* Aldershot: Ashgate, 2015.

National Academies. *Disaster Resilience: A National Imperative.* Washington: The National Academies, 2012.

Norris, F.H., S.P. Stevens, B. Pfefferbaum, K.F. Wyche and R.L. Pfefferbaum. Community resilience as a metaphor, theory, set of capacities, and strategy for disaster readiness. *American Journal of Community Psychology* 41, 2008: 127–150.

Nowlin, M.C. Theories of the policy process: State of the research and emerging trends. *Policy Studies Journal* 39, 2011: 41–60. doi: 10.1111/j.1541-0072.2010.00389_4.x.

Ostrom, E. Public entrepreneurship: A case study in ground water basin management. Doctoral dissertation, Los Angeles: University of California, 1965. Accessed 15 August, 2014. http://dlc.dlib.indiana.edu/dlc/handle/10535/3581?show=full.

Pendall, R., K.A. Foster and M. Cowell. Resilience and regions: Building understanding of the metaphor. *Cambridge Journal of Regions, Economy and Society* 3(1), 2010: 71–84. doi: 10.1093/cjres/rsp028.

Petridou, E. Theories of the policy process: Contemporary scholarship and future directions. *Policy Studies Journal* 42, 2014: S12–S32. doi: 10.1111/psj.12054.

Petridou, E., I. Narbutaité-Aflaki and L. Miles. Unpacking the theoretical boxes of political entrepreneurship. In: I. Narbutaité-Aflaki, E. Petridou and L. Miles (Eds.), *Entrepreneurship in the Polis: Understanding Political Entrepreneurship*. Aldershot: Ashgate, 2015, pp. 1–16.

Prindle, D.F. Importing concepts from biology into political science: The case of punctuated equilibrium. *Policy Studies Journal* 40(1), 2012: 21–44. doi: 10.1111/ j.1541-0072. 2011.00432.x.

Schneider, M., P. Teske and M. Mintrom. *Public Entrepreneurs: Agents for Change in American Government*. Princeton: Princeton University Press, 1995.

Sheingate, A.D. Political entrepreneurship, institutional change, and American political development. *Studies in American Political Development* 17, 2003: 185–203.

Simmie, J. and R. Martin. The economic resilience of regions: Towards an evolutionary approach. *Cambridge Journal of Regions, Economy and Society* 3(1), 2010: 27–43. doi: 10.1093/cjres/rsp029.

Smith, D. Business (not) as usual: Crisis management, service interruption and the vulnerability of organizations. *Journal of Service Marketing* 19, 2005: 309–320.

Smith, D. Crisis management: Practice in search of a paradigm. In: D. Smith and D. Elliot (Eds.), *Key Readings in Crisis Management: Systems and Structures for Prevention and Recovery*. London: Routledge, 2006, pp. 1–12.

't Hart, P. and A. Boin. Between crisis and normalcy: The long shadow of post-crisis politics. In: U. Rosenthal, R.A. Boin and L.K. Comfort (Eds.), *Managing Crises: Threats, Dilemmas, Opportunities*. Springfield: Charles C. Thomas, 2001, pp. 28–46.

Walker, B. and D. Salt. *Resilience Thinking: Sustainable Ecosystems and People in a Changing World*. Washington: Island Press, 2006.

Walker, J. and M. Cooper. Genealogies of resilience: From systems ecology to the political economy of crisis adaptation. *Security Dialogue* 42(2), 2011: 143–160. doi: 10.1177/ 0967010611399616.

Wildavsky, A. *Searching for Safety*. New Brunswick: Transaction Books, 1988.

Wilkenfeld, J., K.J. Young, D.M. Quinn and V. Asal. *Mediating International Crises*. London: Routledge, 2005.

Williams, N., T. Vorley and P.H. Ketikidis. Economic resilience and entrepreneurship: A case study of the Thessaloniki city region. *Local Economy* 28(4), 2013: 399–415. doi: 10.1177/ 0269094213475993.

Zebrowski, C. The nature of resilience. *Resilience* 1(3), 2013: 159–173. doi: 10.1080/21693293. 2013.804672.

Section II

Integration

This section aims to delve deeper into the research on resilience and how its approaches have been integrated into different disciplines. In Chapter 5, Vlajic investigates the issues of vulnerability and robustness in food supply chains. There has been considerable interest in supply chain risk in recent years and this chapter starts by bringing together some of this work. Vlajic continues by showing the importance of redesign strategies on reducing the impact of disturbances and that small and medium-sized enterprises (SMEs) have the resources and capability to do so. In Chapter 6, Adamides and Tsinopoulos consider resilience-related capabilities and how these may have affected the forward integration of some firms during the recent global economic crisis. This is especially pertinent as they have investigated the Greek environment that has been particularly hard hit. The authors show that operations strategy is guided by resilience-related capabilities such as flexibility that includes process technology, diversification and robust implementation of strategy. In Chapter 7, Yang focuses on the informational requirements needed during an emergency response to an event. Yang presents the goal-directed information analysis (GDIA) approach to eliciting relevant and timely information required by emergency first responders during an emergency incident. This is now a proven approach and has already been used to assist in the development of emergency decision support systems for mass evacuation in New Zealand.

Section II

Integration

5 Vulnerability and Robustness of SME Supply Chains
An Empirical Study of Risk and Disturbance Management of Fresh Food Processors in a Developing Market

Jelena V. Vlajic

CONTENTS

5.1 INTRODUCTION

The supply chain literature offers different insights into the vulnerability of small and medium enterprise (SME) supply chains. Some researchers found that SMEs' supply chains are less vulnerable in comparison with the supply chains of large organisations (Wagner and Neshat, 2012), while others found that SMEs' supply chains are more vulnerable in comparison with the supply chains of large organisations (Arend and Wisner, 2005; Thun et al., 2011). Differences in these findings can

3

be explained by the existence of contextual factors that influence results. SMEs are often presented as vulnerable due to limited organisational, financial and human resources (Vaaland and Heide, 2007). Considering that a supply chain is as strong as its weakest link (Kleindorfer and Saad, 2005; Slone et al., 2007), these findings do not actually indicate how vulnerable SMEs' supply chains are. Moreover, the literature does not offer explanations of which contextual factors can influence the level of vulnerability of SMEs' supply chains, nor the capabilities of SMEs to decrease their vulnerability and maintain robust performance.

The objective of this study is to shed light on these issues by investigating the level of vulnerability of SMEs and by analysing to what extent characteristics of 'risky' products and risks related to the business environment contribute to the capability of SMEs' supply chain to manage disturbances. Moreover, additional insights are provided for the type of redesign strategies the SME might select to maintain robust performance and to decrease vulnerability. This research contributes to the supply chain risk management literature with insights on the SME's supply chain vulnerability and its capability to manage disturbance when exposed to the influence of contextual factors that increase risks to SMEs' supply chain.

This chapter is structured as follows. Section 5.2 is devoted to an application of a theoretical framework for achieving robust supply chains to SMEs' supply chain. In this section, contextual factors that might influence the success of an SME to manage disturbance such as the riskiness of the product and riskiness of the business environment are also considered. The employed methodology and the use of an exploratory case study to investigate how these contextual factors contribute to the robustness of the company's performance and supply chain vulnerability while the company manages internal and external vulnerability sources are described in Section 5.3. Sections 5.4 and 5.5 are focused on a brief case description and the main findings related to the company's vulnerability levels in relation to susceptibility to various types of disturbance and the capability of the company to manage these disturbances and their vulnerability sources.

5.2   A THEORETICAL FRAMEWORK

In the supply chain management literature, robustness is considered as the ability of a supply chain to continue to function well even when faced with disturbances in its processes (Dong, 2006; Tang, 2006; Waters, 2007; Stonebraker et al., 2009; Wieland and Wallenburg, 2012; Vlajic et al., 2013).

Generally, to achieve robustness supply chains have to manage disturbances successfully. Vlajic et al. (2012) developed a framework for robust supply chains that considers the main building blocks for successful disturbance management and relations between them (Figure 5.1). The first building block is a supply chain scenario which represents an internally consistent view of a possible instance of the logistics supply chain concept (van der Vorst, 2000). This concept consists of a managed system (physical design of a network, fixed and mobile resources), a managing system (planning, control and coordination of logistic processes), an information system (information and decision support systems) and an organisational structure (tasks, responsibilities and authorities of departments and employees). As a supply

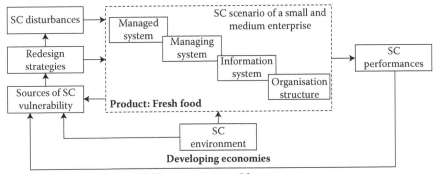

Legend: SC – supply chain **Bold letters** – contextual factors

FIGURE 5.1 Framework for achieving robust supply chains. (Adapted from Vlajic, J.V., van der Vorst, J.G.A.J., Haijema, R., 2012, *International Journal of Production Economics*, 137(1), 176–189.)

chain normally consists of multiple participants, from multinational companies to SMEs, connected by flows of materials, information and finance, analysis of the supply chain scenario is usually centred on the focal company (c.f. Lambert and Cooper, 2000).

The robustness of the supply chain scenario is reflected in its performance. Any deviation of a defined performance range indicates the presence of disturbances in the supply chain process and increased supply chain vulnerability. Disturbances are considered on the process level and detected as any deviation from quantity, quality and time-related specifications.

To manage disturbances successfully, two types of redesign strategies can be used. In this context, redesign strategies are defined as strategic and tactical plans leading to operational actions that aim to reduce the vulnerability of the supply chain by making changes in core elements of the supply chain scenario (Vlajic et al., 2012). The first type of redesign strategy aims to prevent disturbances by avoiding or reducing the probability of exposure to vulnerability sources (Wagner and Bode, 2009). Vulnerability sources may cause disturbances and, in the literature, are categorised as external or internal (Asbjørnslett and Rausand, 1999). External vulnerability sources consider market, societal, infrastructural, financial, legal and environmental factors in a supply chain's business environment. These are considered as uncontrollable from the company's standpoint. Internal vulnerability sources are concerned with the supply chain and the company's perspective of the supply chain scenario (Vlajic, 2012) and are considered as controllable to some extent (Simchi-Levi et al., 2008). The importance of vulnerability sources can be assessed by investigating how often they cause disturbances, the extent to which performances are outside specifications when disturbance occurs, and how easy it is to detect the disturbance and its cause. The second type of redesign strategy aims to mitigate the impact of disturbances by modifying elements of the supply chain scenario, so that impact of disturbances is amortised, for example, by keeping buffer stocks or increasing process flexibility (Wagner and Bode, 2009).

The selection of a particular redesign strategy depends on the characteristics of disturbances and its vulnerability sources (Vlajic et al., 2013). While the majority of papers in the supply chain risk management literature focuses on identification and analysis of risks and disturbances, as well as identification of appropriate or optimal redesign strategies that can be used when disturbances occur, there is little research devoted to opportunities and challenges of SMEs for successful disturbance management.

By comparison with large companies SMEs are perceived as slightly more vulnerable (Thun et al., 2011) for the following reasons: a typically smaller assortment, fewer customers and lower volume, higher capital and transaction costs, a reactive nature in the company's strategies, and the presence of limited resources (Arend and Wisner, 2005). Moreover, SMEs usually have more difficulties managing business processes in the global environment and in overcoming entry and trading barriers (Ritchie and Brindley, 2000). Additionally, they usually operate under conditions of a weaker cash flow and lower equity reserves (Thun et al., 2011). However, SMEs are often characterised by informal exchanges of information between employees (Roebuck et al., 1995), knowledge of local markets, knowledge-based advantages (Arend and Wisner, 2005) and flexible organisational structure (Vaaland and Heide, 2007). Thus, some findings indicate that the lesser complexity of business situations of SMEs makes them less vulnerable than large companies (cf. Wagner and Neshat, 2012).

The literature implies that SMEs have difficulties in managing disturbances, but also that they might have advantages in using certain sets of redesign strategies for managing disturbances caused by internal and external vulnerability sources. Considering specificities of the SMEs' supply chains, it would be important to obtain additional insights into the level of SMEs' supply chains, that is, susceptibility to disturbances in its processes.

Moreover, an SME's vulnerability and robustness of its performance might depend on specificities of vulnerability sources and the capability of companies to implement adequate redesign strategies. These specificities are captured by relevant contextual factors. For this study, characteristics of a product and supply chain business environment are selected as relevant contextual factors. If characterised as risky, both product and supply chain business environment characteristics might contribute to supply chain vulnerability. Considering product characteristics, fresh food products, for example, have a huge 'risk' potential (Henson and Reardon, 2005), due to specific product characteristics (e.g. shelf-life constraints, sensitivity to temperature changes and environmental conditions, biological variations and long production throughput times) and influence operations in logistics processes (van der Vorst, 2000). While the literature recognises risk associated with certain types of products and consequential limitations in the management and organisation of logistics processes, it is not known to what extent characteristics of 'risky' products might restrain possibilities of SMEs for disturbance management and what strategies SMEs might use to manage these disturbances.

Considering supply chain business environment characteristics, the following factors might contribute to supply chain vulnerability (Asbjørnslett and Rausand, 1999): financial (e.g. currency fluctuations), market (e.g. price fluctuations, new products launches and fierce competition), social (e.g. civil disturbance), environmental (e.g. natural disasters), infrastructural (e.g. lack of modern road, rail and port infrastructure) and legal (e.g. lack of regulations or low level of regulatory enforcement). However, even in a risky business environment, companies might achieve robustness of their performances by developing reliable relationships with their supply chain partners (c.f. Fynes et al., 2004). Thus, it would be interesting to investigate to what extent the characteristics of a 'risky' environment might restrain the possibilities of SMEs for disturbance management and what strategies SMEs might use to manage these disturbances.

5.3 METHODOLOGY

Existing theories in the supply chain management literature fail to fully explain the level of exposure of supply chains to vulnerability sources related to the specificities of the products, type of companies involved in the chain and the type of business environment, as well as the mechanisms that can be used to manage consequential disturbances. Thus, an exploratory study is needed to develop pertinent propositions for further inquiry (Yin, 2014). The SME is used as the unit of analysis.

In line with specific characteristics of supply chains that lead to increased vulnerability levels, the following criteria for case selection are used:

1. *The SME in the supply chain:* A medium size dairy processor is selected for the exploratory case. The company's core business is the production of a small, traditional assortment of dairy products. The company has a regional supply network and it delivers dairy products to customers in the region. During the selection process, it was identified that the company experiences multiple disturbances during the year.
2. *Risky products*: Perishable food products are considered as high risks in the literature. These kinds of products affect design and realisation of logistics processes – e.g. they limit inventory levels, require frequent testing of product quality and monitoring of production and logistics activities, and, if not managed properly, could create significant costs due to waste of products.
3. *Risky market:* The chosen SME serves a market in a transition country in the Western Balkans and has a total of around 200 competitors. Transition countries are considered as developing markets, and often as risky business environments for investments (OECD, 2012), as well as having particular infrastructural and legal vulnerability sources. For example, logistics performance indicators (Arvis et al., 2012) highlight challenges with regard to efficiency of customs and clearance processes, quality of trade and transport-related infrastructure, ease of arranging competitively priced shipments, competence and quality of logistics services.

Data collection was conducted in 2011. To comply with requirements for data triangulation, data collection was performed by using semi-structured interviews, a field visit, and an inspection of the company's public documents. Interviews were conducted with the company director, the financial director and the quality control technician. Each interview lasted between one and one and a half hours and they were conducted in the factory. The field visit looked at the milk processing operation, the quality of laboratory food safety testing and the factory's distribution storage. The public documents used were the company's newsletters, reports from quality control agencies, and reports from the National Chamber of Commerce relating to the company's performance.

The purpose of data collection was to collect information about disturbances that occur in the company, the types of vulnerability sources and how they impact on the company's performance and the type of redesign strategies the company uses to manage such disturbances. Data collection was based on the case study protocol and consisted of semi-structured questions grouped in the following blocks: general data about the company; data about disturbances and their impact on company performance; data about vulnerability sources; and data about redesign strategies used to manage disturbances. In addition to the narrative that developed from the questions, interviewees were asked to estimate the severity of disturbances and the impact of various vulnerability sources on the company's performance. Estimation was based on principles of process failure modes and effects analysis (PFMEA), that is, each identified disturbance and corresponding vulnerability sources were rated from 1 to 7 by each interviewee (e.g. frequency scale: 1 – never, 2 – very rare ... 6 – very frequently and 7 – always) with regard to the frequency of occurrence, impact on performances and ease of detection.

Because the framework and case study protocol are based on selected variables, it was convenient to transform the gathered data to an effect matrix, which focused on dependent variables (Miles and Huberman, 1994). Here, the independent variable is a vulnerability source, the dependent variable is disturbance and the intervening variable is redesign strategy. The effect matrix enables the analysis and summation of numerical values which indicate what kind and how many vulnerability sources are present in the case study, as well as what kind and how many vulnerability sources can trigger one disturbance and which and how many different redesign strategies can be used in each instance. Moreover, the estimation of vulnerability levels and a ranking of disturbances are based on the use of disturbance priority number (DPN) values. Due to the emphasis on disturbances, the term disturbance priority number is used rather than risk priority number. According to the PFMEA, a DPN can be estimated as the product of the frequency, impact and detectability of a particular disturbance (Scipioni et al., 2002). A DPN value is used as a measure to rank disturbances and their causes, as well as to estimate the vulnerability level of the company. As each of the three dimensions is evaluated on a (integer) scale of 1–7 the DPN takes values of at least 1 and at most $7^3 = 343$. The higher the DPN value, the higher the priority of disturbance or vulnerability source is. Vulnerability levels can be estimated in various ways. In this chapter, vulnerability levels are defined in relation to qualitative assessment of dimensions, that is, negligible: [1–8], low: (8–27], low–medium: (27–64], medium: (64–125], medium–high: (125–216] and high: (216–343].

5.4   CASE DESCRIPTION

Considering the number of employees and annual turnover, the milk processing company is classified as a medium size company. In the following sections, a case description is provided by using elements of the supply chain scenario.

5.4.1 MANAGED SYSTEM

The company consists of the factory with specialised equipment for milk processing, cold storage for raw milk and milk products, distribution fleet, laboratory and administrative offices. The company also owns several small retail outlets in the region. The company produces three groups of products: fresh milk, plain yogurt and cheese. Fresh milk is the most perishable product (5 days of shelf life), while cheese has a longer shelf life (soft cheese up to 5 weeks and hard cheese up to several months). All products are preserved in protective packaging made of carton and plastic materials.

The company delivers its products to 2000 customers consisting of a small number of large retailers and a large number of small, privately owned retail outlets and food service companies. The company sources raw milk from around 800 suppliers.

5.4.2 MANAGING SYSTEM

The company processes 40,000 L of raw milk per day on average. The production plan and product assortment is based on consideration of the demand of large customers (annual contracts) and a forecast of the demand of small customers. Due to oscillations in sourcing of raw milk and the inability to store raw milk, the company sources larger quantities than is needed for daily production. After the fulfilment of planned production, the remaining raw milk is used for production of cheeses (less perishable products). Part of the managing system is related to supplier management, that is, the company makes planned visits to their suppliers to monitor their work (number of animals, their health, breeding plans, etc.). The company is also engaged in customer management through promotional activities, such as participation in national competitions focused on product quality.

5.4.3 INFORMATION SYSTEM

The company uses spreadsheets in combination with databases and customised accounting software to support its business operations. These databases consist of modules that store information about customers and their orders, suppliers and their capacities, as well as production capacities and schedules. Information flows are twofold: formal (by e-mails, documentation) and informal (exchange of information in personal conversations).

5.4.4 ORGANISATIONAL STRUCTURE

The company consists of a few senior staff whose duties are to plan and organise sourcing, production and sales, keep contact with suppliers and customers, perform financial and accounting activities as well as plan and organise distribution. Staff duties and responsibilities are not strictly divided. Alongside the administrative staff the company employs a few workers in production, a full-time certified quality control laboratory manager and a part-time lab technician.

5.5   FINDINGS AND DISCUSSION

In this section key findings are reported followed by an analytical discussion. First, findings related to the vulnerability level of the milk processor's (SME's) supply chain are presented and subsequently more details and analysis of the observed disturbances, their sources and the capabilities of the milk processor to manage them are discussed.

Results of the PFMEA method applied to the case show that the highest vulnerability level experienced by the milk processor is medium–high (DPN = 144, Figure 5.2). Though impact is high and detectability of this disturbance is difficult due to the delays in reporting by the customer, this disturbance happens occasionally indicating that the milk processor is exposed to this vulnerability source occasionally or that the milk processor has mechanisms to manage this disturbance, which requires further analysis and discussion. To investigate this, more detailed analysis is presented starting with the consideration of disturbances caused by internal, company-related vulnerability sources.

Though Wagner and Neshat (2012) found that SMEs are less vulnerable than large companies, their findings indicate that SMEs are vulnerable. Moreover, they found that companies experience increased vulnerability when producing large series. The milk processor is characterised by production of large volumes and relatively narrow variety, which corresponds to mass production or production of large batches (Slack et al., 2013).

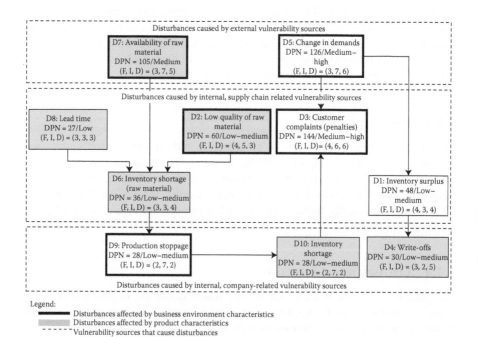

FIGURE 5.2   Relation between disturbances (DPN – disturbance priority numbers are estimated independently for each disturbance).

In combination with their input and output of perishable, 'risky' food products (Henson and Reardon, 2005), the milk processor becomes susceptible to disturbances that result from internal, company-related vulnerability sources (grey fields in Figure 5.2). For example, many milk products are temperature sensitive and have a short expiration date. Thus, preservation of the right quality of milk products requires specialised, often complex production, storage and transport equipment. Any failure in specialised production equipment can cause large disruptions, not only in quantity produced, but it can also have a detrimental effect on the quality of products and on-time deliveries. Moreover, the cost of investment in specialised equipment, especially in the volatile financial markets experienced in many developing countries, might increase the vulnerability of the company. Additionally, the company often has to manage write-offs due to occasional inventory surplus or returned products. Thus, the following proposition can be stated:

P1. If riskiness of the products characterise internal, company-related vulnerability sources, then the manufacturing SME's susceptibility to disturbances is likely increased.

Analysis of DPNs for internal, company-related vulnerability sources indicates that the milk processor has a low–medium vulnerability level, that is, it does not suffer from serious disturbances. Further investigation indicated that the managerial team took a proactive approach to disturbance management even from the design stage, that is, from thorough preparation and planning of the facility design, via careful budget planning, to planning responses in the case of equipment failure, inventory shortage of final products or products in the inventory that are close to the best-before date. Moreover, as the milk processor uses specialised, foreign manufactured equipment, equipment failure might be a rather serious issue due to uncertain availability of spare parts and knowledge for maintenance of this equipment (Huiskonen, 2001), especially in the context of a developing market (Taylor, 1994). Aiming to reduce the frequency of disturbances, the milk processor implemented a set of preventive strategies such as choice of reliable production equipment, organisation of regular maintenance, analysis of customer demands to drive production planning and reduce piling of inventory of final products and consequential write-offs. The company would appear to make the right decisions in regard to these issues as according to the financial manager: '*In last eleven years, the company had to stop production due to equipment failure only once*'. Thus, use of all these preventive strategies contributes to the stability of the production system and robustness of its performance. The use of impact reduction redesign strategies is somewhat limited due to product and business environment characteristics. For example, due to the perishable nature of many milk products the milk processor cannot keep large safety stocks of final products and raw materials as insurance against failures of supply or production. Also, to shorten repair times, the producer stocks important spare parts and maintains contacts with certified maintenance teams in the region as well as abroad.

P2. In the presence of product-related risks, preventive redesign strategies are likely used by a manufacturing SME to manage disturbances caused by internal, company-related vulnerability sources. Riskiness of the business environment might affect the manufacturing SME's capability to successfully manage disturbances.

The milk processor's supply chain consists of many small farms (suppliers) and two main suppliers for packing material on the one side, and multiple large retail customers and a large number of smaller customers on the other side. Thus, the supply and the demand side of its supply chain are analysed separately.

Looking at the supply side, farms produce small quantities of milk and the quality of the milk might vary due to sickness of animals, variations in quality of animal feed, seasonal variation, etc. Due to the high perishability of raw milk, it is not possible to keep a buffer that would shield production. For the same reason, lead time has to be short and as a consequence the milk processor has only local suppliers. The low quality of raw milk can also be traced to external vulnerability sources in a supply chain where small farmers are not educated sufficiently on how to prevent sickness of animals during various seasons, how to plan animal breeding, what feed to use and how to increase farm productivity. For instance, the company director states: *'Low milk production is a consequence of traditional feed choices and use of traditional cow breeds on the farm'*. The small size of many farms does not lend itself to high levels of development that might be technically challenging and relatively costly for a small enterprise.

Thus, due to the variability of quantity and quality of suppliers' inputs (van der Vorst, 2000), as well as constraints in the supply base (Giunipero and Eltantawy, 2004) and characteristics of suppliers (Zsidisin, 2003), companies might also become susceptible to disturbances on the supply side that result from a combination of internal, supply chain-related vulnerability sources and riskiness of a product.

Looking at the demand side, large customers (referred to by informants as *'powerful retail systems'*) usually order large quantities of products, have strict specifications about delivery times, prefer to have a large assortment of products in their retail outlets, as well as to organise promotions and present new types of milk products. Customer complaints are considered (*'measured'*) through the return of unsold products and claims due to deliveries being out of specification (e.g. delivered quantity less than ordered and delayed delivery). This disturbance happens a few times per year, has a high impact on the company's financial performance (in particular, when the company has to pay penalties or to take back products) and is difficult to detect early because penalties are often visible only in end-of-the-month reports. More specifically, according to the financial director, *'Large customers demand use of quantity-flexibility contracts'*. Quantity-flexibility contracts are risky for a supplier, because the supplier is obliged to provide a full refund for returned (unsold) items as long as the number of returns is not larger than a certain quantity (Simchi-Levi et al., 2008). Typically, the supplier is also obliged to organise and meet the cost of the return of unsold items. Looking at this issue in more detail, field data indicates that large customers dictate details of the contracts, that is, that power imbalance exists (Matopoulos et al., 2007). As the company director states: *'They [large customers] set the rule for quantity of products in return, as well as for quality parameters. We have to accept that because they cover large percent of market ...'*. Large customers also define tolerance specification for delivery and stock-out in their premises, as the financial director states: *'If they [large customers] have large demand and our products are sold, we have to deliver new batch in very short period of time'*. Complaints by large customers that products are out of specification and the resultant product returns give rise to the highest priority disturbance for the milk processor. This is not surprising since powerful retail systems often transfer risk to their suppliers (Peck, 2005) and even impose new systems (e.g. specific types of labelling and packaging) on weaker business partners (Coe and Hess, 2005). This situation is especially present in developing markets, which are characterised by low supply chain transparency (Roth et al., 2008), the struggle to comply with food safety standards due to resource limitations (Henson and Humphrey, 2010), incompatibilities of production and marketing systems (Donovan et al., 2001), an underdeveloped and/or fragmented retail sector and the large number of intermediaries (Lorentz et al., 2013), and food traceability issues (Humphrey, 2007).

Based on the analysis of the supply and demand side of the milk processor's supply chain, the following proposition is developed:

P3. When supply and demand-related vulnerability sources are characterised by riskiness of the product and riskiness of the business environment, then the manufacturing SME's susceptibility to disturbances is likely increased.

Low availability or low quality of raw milk can cause production stoppages and inventory shortages of final products due to the inability of the company to keep a buffer of raw milk. Though the use of impact reduction strategies is again somewhat limited by product characteristics, there are still a number of them that can be applied. For example, to reduce impact of disturbances in the case of supplier failure, the company orders slightly higher quantities of raw milk than planned, which results in high utilisation of production capacity (c.f. Vlajic et al., 2013). In this way, variation in input does not affect required production output, and the company gets increased flexibility in realisation of the production program. However, this strategy leads to occasional accumulation of the inventory of cheese, a less perishable product which is produced from the surplus of raw milk. In the rare cases of larger disruption in sourcing, the company has in place arrangements to acquire additional quantities of milk from other processors who have a surplus, that is, emergency transhipment (Simchi-Levi et al., 2008). Moreover, the regular use of their own transport resources and the use of a third-party logistics provider (3PL) when needed result in short lead times for emergency deliveries. Looking at the use of preventive strategies, it is noticed that their use is also limited due to the restricted resources of the milk processor. '*Though our purchasing team has frequent contact with our suppliers, and we belong to the same cooperative, we are not capable of helping them to solve problems at their farms more than giving advice and recommendations*', states the financial director, which implies low ability of the SME to participate in supplier development and develop their network (Singh et al., 2008). Also, to eliminate possibilities of contamination of high-quality milk with low-quality milk, the company conducts quality control tests on the farms. This allows the SME to detect this disturbance effectively, although it remains difficult to identify its causes and to prevent it.

The analysis of DPNs for internal, supply chain-related vulnerability sources indicates that the milk processor suffers from serious disturbances caused by demand side vulnerability sources. Customer complaints are considered as disturbances related to deviations in quantity or quality of products delivered, and this disturbance causes medium–high vulnerability level and thus has the highest priority.

However, the milk processor experiences low–medium vulnerability with respect to inventory surplus of final products and product write-offs because of the redesign strategies used. Collected data indicate that the milk processor relies again on impact reduction redesign strategies to manage frequent customer complaints with regard to the assortment and changing preferences for milk products by large customers, as well as inventory surplus. The main reason lies in the fact that complex, specialised equipment for milk processing limits the possibilities to increase the variety of final products. Thus, the company is not able to satisfy such demands without costly investment in additional equipment and diversification of customers represents the dominant strategy to manage disturbances (Waters, 2007). Moreover, by keeping a diverse customer base, returned products can be sold to customers that have broader tolerance specifications (e.g. processors of pet and domestic animal feeds). Return flows are combined with distribution tasks by using their own fleet, which is considered as a good, sustainable logistics practice (Tibben-Lembke and Rogers, 2002; Kleindorfer et al., 2005).

Based on the analysis of redesign strategies used in the supply and the demand side of the milk processor's supply chain, the following proposition can be stated:

P4. In the presence of product-related risks, impact reduction redesign strategies are likely used by the manufacturing SME to manage disturbances caused by internal, supply-related vulnerability sources, while in the presence of business environment-related risks, these strategies are likely used by the manufacturing SME to manage disturbances caused by internal, demand-related vulnerability sources.

Further analysis of the vulnerability sources reveals a number of external vulnerability sources that represent the root causes of some disturbances discussed above. Field data indicate that the real threat of raw material inventory shortages results from the low availability of raw material on the local market. The challenge here is that it is difficult for the milk processor to detect external vulnerability sources, as well as to reduce the frequency and the impact of the disturbances they create. More specifically, the threat is compounded by the poor ability of the control and planning system of the company to identify and manage this disturbance to their supply market. Though low availability of raw material on the local market occurs once every few years, it has a huge impact because it affects all companies in the sector. Sector analysis in this developing market revealed that the supply base of the local market is shrinking as small farmers are leaving the business. The main reason for this lies in low milk prices on the national market, as well as delays in cash flow. Moreover, suppliers also speculate who to select as a favourable customer and when to leave milk production in favour of other farm production such as cattle rearing. Thus, climate changes that affect feed quality, variability and seasonal peaks in raw material quality (van der Vorst, 2000; Matopoulos et al., 2007), the fluctuation of the price of raw material (Matopoulos et al., 2007; Trkman and McCormack, 2009), the low level of the supplier's education related to production and farm management, a centralised state that lacks the flexibility to rapidly adapt its regulations to market changes, poor financial systems and contractual discipline and enforcement (van Veen, 2005) represent the main external factors that contribute to disturbances in the SME's supply chains.

P5. Riskiness of the product and the business environment characterise external vulnerability sources that likely increase the manufacturing SME's susceptibility to disturbances.

The impact of these vulnerability sources, however, is not high due to the combination of the specificities of business environment characteristics and the use of impact reduction redesign strategies. For instance, in spite of government interventions in setting market prices and regulating competition in the dairy industry, as well as poor incentives and only a partially developed program for industry development, the company is reducing its vulnerability by exploiting knowledge of the local supply and demand market. For example, it uses close communication with selected local suppliers and a simplified form of tracking and tracing to provide insights into its regional supply market that enables it to predict both raw material quality and availability. Thus, by building communication, relationship and trust with suppliers (Matopoulos et al., 2007; Dani and Deep, 2010) and by developing knowledge about the industry, the company is increasing chances for timely reaction in the case of disturbances caused by external factors.

P6. In the presence of both, product- and business environment-related risks, impact reduction redesign strategies are likely used to manage disturbances caused by external vulnerability sources.

5.6   CONCLUSIONS, LIMITATIONS AND FURTHER RESEARCH

The supply chain risk management literature offers partial and conflicting findings in relation to the vulnerability of SME supply chains. It is highly likely that certain contextual factors play a significant role in successful disturbance management and the resulting vulnerability levels. The aim of this chapter was to investigate an SME's susceptibility to disturbances, the vulnerability levels and the SME's ability to manage disturbances by considering relevant contextual factors.

The SME's supply chain is likely to be more susceptible to disturbances in the presence of contextual factors that carry certain levels of risk. Supply chains are product and company specific, (Reiner and Trcka, 2004), and details of their logistics and production processes depend on the business environment in which they operate (Trkman and McCormack, 2009). Thus, a 'risky' product and business environment can be indicative of an SME's susceptibility to disturbance and should raise questions about its ability to manage them.

We base these findings on an exploratory study of an SME's fresh food supply chain that operates in a developing market. These conclusions suggest that fresh food supply chains of SMEs in developing markets are prone to disturbances of their logistics and production processes due to the 'riskiness' of fresh food products and the 'riskiness' of a developing market. Moreover, the SMEs are also characterised by risky features, that is, limited organisational, financial and human resources that might influence their capability to manage disturbances. As a consequence, the SME manages disturbances by considering limitations imposed by these contextual factors and by choosing adequate redesign strategies, more specifically as follows:

- Riskiness of the products characterise internal, company-related vulnerability sources that likely contribute to the SME's susceptibility to disturbances, while riskiness of the business environment might affect its capability to successfully manage disturbances. Manufacturing SMEs likely use preventive redesign strategies to manage disturbances that result from internal, company-related vulnerability sources.
- Both riskiness of the products and riskiness of the business environment characterise internal, supply chain-related vulnerability sources that likely contribute to the SME's susceptibility to disturbances. However, there is strong indication that the riskiness of its product affects more the supply side of the SME's supply chain, while riskiness of the business environment affects more the demand side of the SME's supply chain. The SME's ability to use preventive strategies to manage disturbances is limited, and it is likely that the SME uses impact reduction redesign strategies to manage these kind of disturbances.
- Both the riskiness of the products and the riskiness of the business environment characterise external vulnerability sources that likely contribute to the SME's susceptibility to disturbances. The SME likely manages disturbances by using impact reduction redesign strategies due to limited ability to use preventive strategies.

Thus, the findings show that though SMEs are susceptible to disturbances, this does not necessarily lead to the vulnerability of the SME and its supply chain if the right redesign strategies are implemented.

To conclude, the riskiness of the product and business environment characterise vulnerability sources and thus contribute to the SME's susceptibility to disturbances. Moreover, the findings indicate that these contextual factors might limit the use of preventive redesign strategies. For example, disturbances that result from vulnerability sources characterised by riskiness of the business environment are managed by using impact reduction strategies.

Therefore, it might be interesting to analyse whether this conclusion would change if this type of analysis is performed on a large company, which has more resources and power in the supply chain (Matopoulos et al., 2007). Disturbances that result from vulnerability sources characterised by the riskiness of the product are managed by using both preventive and impact reduction strategies. Similarly, it might be interesting to analyse whether use of preventive redesign strategies would increase if this analysis is performed on a large company that can assess crucial resources in a supply chain and dominate in the supply chain (Cox, 1999).

Since the data is limited to one case study, it is difficult to estimate what contextual factors dominate in contribution to the SME's susceptibility to disturbances. However, supply-related vulnerability sources tend to be characterised more by product-related risks, while demand-related vulnerability sources tend to be characterised more by business environment-related risks.

The findings and conclusions are the result of a single exploratory case study and some statements might be industry specific. To generalise from these findings and build a theory of the vulnerability of SMEs' supply chains and robustness of their performance, it is necessary to test the stated propositions on other case studies. Moreover, an investigation of vulnerability levels is based on subjective estimation of frequency, impact and detectability of disturbances in an SME's supply chain with due consideration given to certain redesign strategies. The development of a simulation model could be helpful in investigating the effectiveness of certain redesign strategies in observed supply chains.

REFERENCES

Arend, R.J., Wisner, J.D., 2005, Small business and supply chain management: Is there a fit? *Journal of Business Venturing*, 20, 403–436.

Arvis, J.F., Mustra, M.A., Ojala, L., Shepherd, B., Saslavsky, D., 2012, Connecting to compete 2012, trade logistics in the global economy, the logistics performance index and its indicators, *The International Bank for Reconstruction and Development*, The World Bank, Washington.

Asbjørnslett, B.E., Rausand, M., 1999, Assess the vulnerability of your production system, *Production Planning and Control*, 10(3), 219–229.

Coe, N.M., Hess, M., 2005, The internationalization of retailing: Implications for supply network restructuring in East Asia and Eastern Europe, *Journal of Economic Geography*, 5, 449–473.

Cox, A., 1999, Power, value, and supply chain management, *Supply Chain Management: An International Journal*, 4(4), 167–175.

Dani, S., Deep, A., 2010, Fragile food supply chains: Reacting to risks, *International Journal of Logistics Research and Applications: A Leading Journal of Supply Chain Management*, 13(5), 395–410.

Dong, M., 2006, Development of supply chain network robustness index, *International Journal of Services Operations and Informatics*, 1(1/2), 54–66.

Donovan, J.A., Caswell, J.A., Salay, E., 2001, The effect of stricter foreign regulations on food safety levels in developing countries: A study of Brazil, *Review of Agricultural Economics*, 23(1), 163–175.

Fynes, B., de Burca, S., Marshall, D., 2004, Environmental uncertainty, supply chain relationship quality and performance, *Journal of Purchasing and Supply Management*, 10, 179–190.

Giunipero, L.C., Eltantawy, R.A., 2004, Securing the upstream supply chain: A risk management approach, *International Journal of Physical Distribution and Logistics Management*, 34(9), 698–713.

Henson, S., Humphrey, J., 2010, Understanding the complexities of private standards in global agri-food chains as they impact developing countries, *Journal of Development Studies*, 46(9), 1628–1646.

Henson, S., Reardon, T., 2005, Private agri-food standards: Implications for food policy and the agri-food system, *Food Policy*, 30, 241–253.

Huiskonen, J., 2001, Maintenance spare parts logistics: Special characteristics and strategic choices, *International Journal of Production Economics*, 71(1–3), 125–133.

Humphrey, J., 2007, The supermarket revolution in developing countries: Tidal wave or tough competitive struggle, *Journal of Economic Geography*, 7, 433–450.

Kleindorfer, P.R., Saad, G.H., 2005, Managing disruption risks in supply chains, *Production and Operations Management*, 14(1), 53–68.

Kleindorfer, P.R., Singhal, K., van Wassenhove, L.N., 2005, Sustainable operations management, *Production and Operations Management*, 14(4), 482–492.

Lambert, D.M., Cooper, M.C., 2000, Issues in supply chain management, *Industrial Marketing Management*, 29, 65–83.

Lorentz, H., Kittipanya-ngam, P., Srai, J.S., 2013, Emerging market characteristics and supply network adjustments in internationalizing food supply chains, *International Journal of Production Economics*, 145, 220–232.

Matopoulos, A., Vlachopoulou, M., Manthou, V., Manos, B., 2007, A conceptual framework for supply chain collaboration: Empirical evidence from the agri-food industry, *Supply Chain Management: An International Journal*, 12(3), 177–186.

Miles, M.B., Huberman, A.M., 1994, *Qualitative Data Analysis,* II edition, Sage Publications, Thousand Oaks, CA.

OECD, 2012, *Competitiveness and Private Sector Development: Eastern Europe and South Caucasus 2011: Competitiveness Outlook*, OECD Publishing, Paris.

Peck, H., 2005, The drivers of supply chain vulnerability: An integrated framework, *International Journal of Physical Distribution and Logistics Management*, 35(4), 210–232.

Reiner, G., Trcka, M., 2004, Customized supply chain design: Problems and alternatives for a production company in the food industry, A simulation based analysis, *International Journal of Production Economics*, 89, 217–229.

Ritchie, B., Brindley, C., 2000, Disintermediation, disintegration and risk in the SME global supply chain, *Management Decision*, 38(8), 575–583.

Roebuck, D.B., Sightler, K.W., Brush, C.C., 1995, Organizational size, company type, and position effects on the perceived importance of oral and written communication skills, *Journal of Managerial Issues*, 7(1), 99–115.

Roth, A.V., Tsay, A.A., Pullman, M.E., Gray, J.V., 2008, Unravelling the food supply chain: Strategic insights from China and the 2007 recalls, *Journal of Supply Chain Management*, 44(1), 22–39.

Scipioni, A., Saccarola, G., Centazzo, A., Arena, F., 2002, FMEA methodology design, implementation and integration with HACCP system in a food company, *Food Control*, 13, 495–501.

Simchi-Levi, D., Kaminsky, P., Simchi-Levi, E., 2008, *Designing and Managing the Supply Chain, Concepts, Strategies, and Case Studies*, III edition, McGraw-Hill Irwin, Boston.

Singh, R.K., Garg, S.K., Deshmukh, S.G., 2008, Strategy development by SMEs for competitiveness: A review, *Benchmarking: An International Journal*, 15(5), 525–547.

Slack, N., Brandon-Jones, A., Johnston, R., 2013, *Operations Management*, VII edition, Pearson, London, UK.

Slone, R.E., Mentzer, J.T., Dittmann, J.P., 2007, Are you the weakest link in your company's supply chain? *Harvard Business Review*, 85(9), 116–127.

Stonebraker, P., Goldhar, J., Nassos, G., 2009, Weak links in the supply chain: Measuring fragility and sustainability, *Journal of Manufacturing Technology Management*, 20(2), 161–177.

Tang, C.S., 2006. Robust strategies for mitigating supply chain disruptions, *International Journal of Logistics: Research and Applications*, 9(1), 33–45.

Taylor, D.H., 1994, Problems of food supply chain logistics in Russia and the CIS, *International Journal of Physical Distribution and Logistics Management*, 24(2), 15–22.

Tibben-Lembke, R.S., Rogers, D.S., 2002, Differences between forward and reverse logistics in a retail environment, *Supply Chain Management: An International Journal*, 7(5), 271–282.

Thun, J.-H., Drüke, M., Hoenig, D., 2011, Managing uncertainty – An empirical analysis of supply chain risk management in small and medium-sized enterprises, *International Journal of Production Research*, 49(18), 5511–5525.

Trkman, P., McCormack, K., 2009, Supply chain risk in turbulent environments – A conceptual model for managing supply chain network risk, *International Journal of Production Economics*, 119, 247–258.

Vaaland, T.I., Heide, M., 2007, Can the SME survive the supply chain challenges, *Supply Chain Management, An International Journal*, 12(1), 20–31.

Van der Vorst, J.G.A.J., 2000, Effective food supply chains: Generating, modelling and evaluating supply chain scenarios, Doctoral dissertation, Wageningen University, Wageningen, the Netherlands.

Van Veen, T.W.S., 2005, International trade and food safety in developing countries, *Food Control*, 16, 491–496.

Vlajic, J.V., 2012, Robust food supply chains: An integrated framework for vulnerability assessment and disturbance management, Doctoral thesis, Wageningen University, the Netherlands.

Vlajic, J.V., van der Vorst, J.G.A.J., Haijema, R., 2012, Framework for designing robust supply chains, *International Journal of Production Economics*, 137(1), 176–189.

Vlajic, J.V., van Lokven, S.W.M., Haijema, R., van der Vorst, J.G.A.J., 2013, Using vulnerability performance indicators to attain food supply chain robustness, *Production Planning and Control*, 24(8/9), 785–799.

Wagner, S.M., Bode, C., 2009, Dominant risks and risk management practices in supply chains, in: Zsidisin, G.A., Ritchie, B. (Eds.), *Supply Chain Risk: A Handbook of Assessment, Management, and Performance*, Springer, New York. pp. 271–290.

Wagner, S.M., Neshat, N., 2012, A comparison of supply chain vulnerability indices for different categories of firms, *International Journal of Production Research*, 50(11), 2877–2891.

Waters, D., 2007, *Supply Chain Risk Management, Vulnerability and Resilience in Logistic*, Kogan Page Limited, London and Philadelphia.

Wieland, A., Wallenburg, C.M., 2012, Dealing with supply chain risks, linking risk management practices and strategies to performances, *International Journal of Physical Distribution and Logistics Management*, 42(10), 887–906.

Yin, R.K., 2014, *Case Study Research, Design and Methods*, V edition, Sage, London.

Zsidisin, G.A., 2003, Managerial perceptions of supply risk, *Journal of Supply Chain Management*, 39(4), 14–26.

6 Survival in Economic Crisis through Forward Integration
The Role of Resilience Capabilities

Emmanuel D. Adamides and Christos Tsinopoulos

CONTENTS

6.1   INTRODUCTION

Despite the ambiguity regarding the exact meaning of the term resilience (resistance, flexibility, reliability or redundancy; Seville et al., 2008), resilience is a sought-after characteristic of organisations of various sizes and scopes in times of crisis (Bhamra et al., 2011). Economic crises have been responsible for significant disruption in the way countries, economies and organisations function. Resilience implies that organisations have the ability and capability to withstand the shocks of economic crises and continue to operate effectively. Hence, building resilience and ensuring a quick return to equilibrium (Bhamra et al., 2011) has been at the centre of government policy, at the national level and strategic analysis at the firm level. At both levels the aim has been to improve the ability to return swiftly to normality. This has attracted much attention in recent years in national and international newspapers and academic journals. In this chapter, we focus on the impact economic crises have on the firm's environment and its strategic management and how the latter is modulated by the existence of resilience-related capabilities. More specifically, we explore which dimensions of built resilience enable firms to adopt strategies of vertical integration for surviving economic crises.

In capitalist economies, economic, financial and fiscal crises are characterised by dramatic changes in macro-economic and social indicators. These are often followed by policies and structural reforms aimed at absorbing the crises' effects and at attracting domestic and international capital. In response to these challenges and institutional pressures, firms have to re-think their structure and redirect their strategic orientation (DiMaggio and Powell, 1983; Hall and Soskice, 2002). Both corporate and functional level strategies need to be modified by adjusting organisational structures and processes accordingly (Grossler et al., 2006; Witcher and Chau, 2012). Changes in the institutional environment are initially experienced at the boundaries of the firm (suppliers, customers) and at the *operations* level. In a realistic strategy formation perspective (Mintzberg, 1994; Burgelman, 2005), these changes provide signals in a bottom-up fashion. As a result, the adjustment of operations strategy proceeds, or is in interplay, with competitive/corporate level strategy formation.

Extant research has found that, in general, alignment between the competitive/corporate and operations strategies influences firm performance (e.g. Hayes and Wheelwright, 1984; Swamidass and Newell, 1987; Rhee and Mehra, 2006), and that institutional pressures are determinants of operations choices, at least as far as the adoption of methods and tools are concerned (Kauppi, 2013). In addition, the successful implementation of such a strategic change necessitates the cost-effective deployment of operations resources and capabilities in the upstream or downstream part of the value chain. Consequently, resilience, which is linked with such a capability, will also depend on coordination of operations resources, including the relationships with agents and institutions at the boundary of the firm, which might have been the outcomes of previous, prior to the crisis, strategies.

In this respect, vertical integration is a strategic move that can be either a corporate level or/and functional level strategic response to institutional pressures that could increase an organisation's resilience to a crisis (Chang, 2003; Chong et al., 2014). Although vertical integration has been out of fashion for some time, it may potentially be a wise strategy in an operational context of institutional transition pressures, uncertainty and low-growth or stagnant markets (Thomson and Strickland, 2001; Osegowitsch and Madhok, 2003). In fact, institutional economists (North, 1990) have argued that changes in the legal and regulatory structures in which the firms operate have an important impact on a firm's choice of its vertical boundaries. Nevertheless, the rejuvenation of vertical integration, especially towards the customer, is motivated by the need to obtain information and develop learning capabilities and/or intangible assets, such as brand name (Ding and Mahbubani, 2013), and not by the need to reduce transaction costs (Osegowitsch and Madhok, 2003). Institutional factors contingent to specific countries determine the importance and function of information flows, knowledge processes and related capabilities and hence influence the choice for integration (Lehrer, 2001).

Regarding institutional pressures stemming from economic crises and the choice of vertical integration, researchers have focused on the role of capabilities in the integration decision in 'abnormal' situations – transitions and reforms – and on the eventual success or failure of the strategy. For instance, by focusing on transactions on the value chain, Brahm and Tarziján (2014) have found that a firm's capabilities strongly mediate the relationship between transaction hazards originating from temporal specificity or an exogenous change and the vertical integration decision. Similarly, Jacobides and Winter (2005), based on an industry level study, have put forward the argument that capabilities co-evolve with transaction costs to set the menu of available choices that firms face in an industry regarding their vertical scope. Diez-Val (2007) has found that firms integrate to create specific investments in capabilities between stages of the value chain, to exploit their pool of knowledge and to guarantee the quality of inputs in the activities of the value chain. Clearly, many of the capabilities discussed in the above literature are related to organisational resilience (Välikangas and Romme, 2012).

In this chapter, we rely on the structural components of the 'varieties of capitalism' perspective of political economy (Hall and Soskice, 2002) to put forward the argument that integration strategies in periods of economic crisis necessitate the use of different facets of organisational resilience and that resilience requires understanding the institutional context. This has been chosen because it is a firm-centred political economy that adopts a relational view of the firm *vis-à-vis* its associated parties, for example, suppliers, customers, collaborators, trade unions, business associations and governmental agencies. Often, these are problematic relations (Milgrom and Roberts, 1992), and as firm capabilities are mostly and ultimately relational, the success of a firm and its strategy depends on its ability to coordinate effectively with a wide range of actors belonging to five institutional spheres: *industrial relations, vocational training* and *education* for inflow of suitable workforce, *corporate governance* that includes the means for accessing finance, *inter-firm relations* (e.g. suppliers and clients) and coordination issues vis-à-vis a *firm's own employees* (e.g. relationships and information sharing) (Hall and Soskice, 2002). Depending on the particular variety, coordination may take place by market forces (liberal market economies), government intervention (coordinated market economies) or a mixture of the two.

It should be noted that, in the chapter, we employ this 'varieties of capitalism' perspective eclectically. We use it as an *instrument for structuring* the information concerning a firm's environment, and not as a means to compare political economies. Our empirical research is based on data from Greece, a country that has been severely hit by financial and fiscal crisis since 2008.

In Section 6.2, we review the link between resilience, vertical integration and the institutional environment in which a firm operates. Then, we explain the dynamics of the specific institutional environment in Greece prior to and after the crisis and present the development of four hypotheses. The hypotheses relate the operation's strategic decision of forward integration with the resilience-related capabilities of understanding sectoral dynamics, strategic flexibility through product and operations diversification, and robustness in strategy implementation through the implementation of process technology. Then, we present the analysis of the results of the empirical research and summarise the main findings. Finally, we end the chapter by discussing the conclusions and limitations of our work.

6.2   INSTITUTIONAL PERSPECTIVES ON INTEGRATION AND THE ROLE OF RESILIENCE

The integration decision is primarily a strategic decision, which may be part of a proactive (top-down) strategy or a response to operational realities expressed as a supply-chain coordination issue (i.e. in the context of emerging strategy). For instance, the vulnerability of a company's supply chain, caused by the unwillingness, or inability, of specific partners to fully cooperate (Sutcliffe and Zaheer, 1998), may surface as an operational issue, for example, in the form of the 'bullwhip effect' (Christopher and Lee, 2004). This may lead to the decision towards integration, or partial integration (Arya and Mittendorf, 2013), after other 'fire-fighting' operational level measures prove inadequate (Wagner and Bode, 2006).

Operation strategy decisions on vertical integration concern the direction, the extent and the balance among the activities being integrated. At a higher strategic level, integration decisions are related to the acquisition of supplier and customer firms or the development of new ones to accomplish the activities of suppliers and customers. There is a variety of reasons behind integration decisions based on different theoretical foundations. Transaction cost economics (TCE) has provided the most popular reason for integration as it suggests the need for aligning transactions with governance structures in a transaction cost-economising way (Williamson, 1990). Related to TCE is the integration theory, put forward by Grossman and Hart (1986), according to which asset ownership mitigates the fact that contracts cannot account for all future contingencies thus limiting the range of possible uses of assets. Possessing the residual control rights of a firm's assets implies that these assets can be deployed in any way the owner of the assets desires.

In more practical terms, an obvious advantage of, and an incentive for, integration comes from the elimination of double marginalisation (Cachon, 2003; Lin et al., 2014) and the achievement of better coordination (Kaya and Özer, 2012). Forward integration allows a producer to better control the distribution of its products, achieve speed and dependability and monitor its prices. More importantly, it can have direct access to market information without the presence of intermediate filters that result in supply-chain coordination problems (Christopher and Lee, 2004). On the other hand, backward integration allows for better control of suppliers and is a suitable strategic move when quality is an operations objective (Richardson, 1996; Diez-Vial, 2007; Lin et al., 2014). In many cases, integration does not result in lower costs as the firm loses flexibility in choosing the most cost-effective supplier and/or the most cost-effective distributor (Ding and Mahbubani, 2013). Nevertheless, it can control its overall long-term efficiencies (Frohlich and Westbrook, 2001) by reducing time delays, improving learning by doing, protecting organisational competencies, avoiding difficulties in relationship management and secrecy of operations. In addition, vertical integration permits faster recovery after a disruption (Hendricks et al., 2009) and secures distribution channels in markets with increased uncertainties (Rangan et al., 1993; Brahm and Tarziján, 2014). Naturally, when organisations integrate forward, they need to change their focus from operational excellence to customer commitment (Wise and Baumgartner, 1999).

There have been a number of different studies, based on different theoretical backgrounds, examining when to integrate, in which direction and in which form (e.g. Koufteros et al., 2007; Flynn et al., 2010; Guan and Rehme, 2012; Ding and Mahbubani, 2013). Although many arrive at somewhat contradictory results, they all agree on the importance of contextual factors that influence the integration decision. Of particular importance to our study is the literature that relates environmental uncertainty and institutional changes stemming from economic crises to the integration decision. For Sutcliffe and Zaher (1998), different sources of uncertainty have different implications for vertical integration. Strategic uncertainty related to participants of the supply chain is a motivating factor for integration. On the other hand, the marketing literature provides weak support for the relation between the degree (or likelihood) of forward integration and environmental uncertainty. Asset specificity and volume uncertainty make this relationship stronger (Walker and Weber, 1984; Heide and Stump, 1995; Sutcliffe and Zaher, 1998; David and Han, 2004; Nagurney et al., 2005; Cadeaux and Ng, 2012).

Related to financial, fiscal and economic crises are institutional changes, such as market liberalisation (or regulation), changes in ownership and/or governance structures, which create environmental uncertainty that may lead to strategic re-direction. Toulan (2002), based on a single company study in Argentina, has concluded that market liberalisation leads to shrinking vertical integration. Clearly, this is a finding that relies on the specificities of the particular institutional setting of the specific country and supports the argument that the institutional environment and its evolution as response to crises (Soskice, 2009) influences adoption, or not, of integration strategies.

The effective adoption and implementation of a different operations and/or corporate strategy is an expression of the firm's resilience. As Hamel and Välikangas (2003, p. 53) indicate 'Strategic resilience is not about responding to a one-time crisis. ... It's about continuously anticipating and adjusting ...'. Hence, resilience is an organisational capability associated with culture (Mitroff et al., 1989) and managerial cognitive schemas (Bhamra et al., 2011) which depend on the extent and complexity of strategies and the balance between exploration and exploitation (Volberda, 2003). Resilient people and organisations have the ability to accept reality, have a deep belief and strong values, and the ability to improvise (Coutu, 2002).

Resilient organisations exhibit adaptive capacity (Woods and Wreathall, 2008) that enables them to respond to shocks stemming from their environment and recover fast. Gibson and Tarrant (2010) suggested four strategies that organisations can develop to improve their resilience: resistance, reliability, flexibility and redundancy, flexibility being the most effective in absorbing the shock. In fact, strategic and organisational flexibility has been proposed as the principal requirement for organisations that compete in turbulent environments (Sanchez, 1995; Hayes et al., 2005; Välikangas and Romme, 2012) and is a function of the dynamic capabilities of the firm (Teece et al., 1997).

Before moving to the development of hypotheses and the related empirical research concerning the relation between the operations strategic decision of forward integration and the development of resilience-related capabilities through strategies of flexibility and robustness, we review, under a varieties-of-capitalism perspective, the institutional environment of Greece and the transformations that it underwent as a result of the fiscal, financial and economic crisis and the related stabilisation measures.

6.3 THE SPECIFIC CONTEXT: THE INSTITUTIONAL ENVIRONMENT OF GREECE AND ITS TRANSFORMATION

In the neo-institutional 'varieties of capitalism' school of political economy (Hall and Soskice, 2001; Tylecote and Visintin, 2008), Greece has been traditionally positioned in the 'ambiguous position' (Hall and Soskice, 2001) between the liberal and coordinated market economies of the 'family-state capitalism' variety, or in the Mediterranean model along with other southern European countries (Amable, 2003; Almond, 2011). It is characterised by firms (large, medium and small) owned and managed by families or powerful individuals (Papadakis, 2006) with an important role deserved for the state as producer, customer and regulator. Financial markets have been underdeveloped and the role of financial institutions in financing and participating in large industrial projects has been very limited. There has been strong employment protection, but the participation of employees and their representing organisations in the corporate governance of private firms has been limited as it has not been covered by legislation (workers-participation.eu, 2014). In many sectors, horizontal and/or vertical relations with other firms have been strong, especially towards the supply side (Psychogios and Szamosi, 2007). Lack of industrial policy has kept these relationships on personal terms. On the other hand, relations with banks and other financial institutions have mostly been at arms' length. For many large industries, the state and its agencies have been their major customers. This has resulted in sectoral concentration and (over)regulation of product markets and management decisions (Conway et al., 2006), as well as in relatively high levels of corruption (Ackerman, 2006).

An important characteristic of the Greek variety of capitalism has been the very large percentage of very small independent (family) firms. In Greece, 57% of the working people have been either self-employed or worked in very small companies that employed less than 10 people (in the European Union area this figure is 36%) (Schmiermann, 2008). Clearly, this small size and independence has had its costs: inability to take advantage of economies of scale, very weak innovation activity, difficulty of the state to collect taxes and check for undeclared work and uncertain business continuity after the firm's founders/owners retired. Small size and environmental uncertainty resulted in short termism and foreclosure, as well as unwillingness to cooperate in investing in and managing common resources (Makridakis et al., 1997; Adamides et al., 2003) while small business cooperation has concentrated on agreeing on prices and exercising lobbying in governmental agencies.

The fiscal, financial and economic crises of the late 2000s disturbed the dominant institutional models of weak economies. From the beginning of 2010, the sovereign debt crisis and the signing of the 'Memorandum of Economic and Financial Policies' between the Greek government and the European Commission, the European Central Bank and the International Monetary Fund (the so-called *troika*) resulted in several structural and environmental changes.

The transforming business environment led many Greek firms to seek strategies of survival rather than growth. In this environment of crisis, many small and medium size firms of the retailing and distribution sectors went out of business as they were not able to finance their operations (European Commission, 2013). For some of the larger firms, reforms in the labour market had relatively little effect, mainly because labour-intensive industries, such as the garments sector, had already moved to low labour cost countries with lower wages, whereas others, such as the construction sector, collapsed due to the absence of any domestic demand and their inherent inability to internationalise (Aghion and Roulet, 2011).

Inevitably, the crisis changed the relationships of Greek firms with their surrounding actors, and these changes have been mostly experienced in the practices of everyday operations. Table 6.1 summarises the crisis-induced changes in the varieties-of-capitalism institutional spheres and the pressures/opportunities they entail in the interior and external environment of firms. From the table, it becomes clear that firms needed to modify their operations strategies in order to deal with reduced demand, environmental uncertainty and broken supply chains. This was more so, given the asymmetric nature of the majority of supply chains of Greek industrial firms, as far as the (small) size of supplier and customer organisations and their vulnerability are concerned (Spanos et al., 2001), and taking into account that broken business links were based on personal relations and trust and are difficult to repair.

6.4   CAPABILITIES FOR RESILIENCE AND THEIR RELATION TO THE ADOPTION OF FORWARD INTEGRATION STRATEGIES IN TIMES OF ECONOMIC CRISIS

The decision to vertically integrate is driven mainly by economic and institutional contingencies and, as we explain in this section, enabled by a set of resilience-related capabilities. Forward integration in a period of crisis helps companies to compensate for revenues lost in product sales by undertaking activities such as distribution and maintenance. Furthermore, forward integration helps industrial firms to better control the promotion of their goods and their prices (e.g. flexible pricing for management of inventories). Finally, forward integration allows these organisations to manage their capacity by managing demand instead of absorbing market disturbances by varying their physical capacity.

Using the specific context of the Greek economic crisis, we develop hypotheses relating the operations strategy decision of forward integration with resilience-related organisational capabilities. In particular, we test the relationship between the resilience-related capabilities of building strategic flexibility through product and operations diversification and the implementation of process technology.

TABLE 6.1

The Firms' Institutional Environment in Greece before and after the Crisis

Institutional Sphere	Before the Crisis	After the Crisis	Coordination Pressures/ Opportunities in Value Chain	Coordination Pressures/ Opportunities in Value Chain
Industrial relations	Strong employment protection Weak undeclared labour legislation/ controls	Weaker employment protection Stricter undeclared labour legislation/ controls		Cheaper labour More flexible labour legislation
Vocational training/ education	Specialist, technocratic	Specialist, technocratic	Low job mobility Unemployment	High cost of (re)training
Corporate governance	Family-owned firms Weak financial system Risk aversion by financial institutions	Collapse of financial system	Failure of cash-hungry SMEs-partners in the value/ supply chain	Difficult to finance large-scale projects
Relations with other firms	Based on personal relationship Not supported by industrial policy	Based on personal relationship Not supported by industrial policy	Difficult to find/ trust new partners in the value chain	Implementation of extended operations through technology
Relations with own employees	Centralisation and formalisation of information Unwillingness to delegate responsibility	Centralisation and formalisation of information Unwillingness to delegate responsibility		Formalisation leads to easy implementation of technology-supported processes

6.4.1 UNDERSTANDING SECTORAL DYNAMICS AND INTEGRATION

Our first hypothesis concerns the relationship between the perceived impact of crisis on a sector and the pursuit of a policy of forward integration. Not all sectors are affected in the same way by an economic crisis or undergo the same degree of institutional transformations. Further, not all firms in a sector have the same understanding of the degree of damage made to the sector by the crisis. In consumer sectors, customers prioritise their needs to accommodate their reduced budgets. In the previous crisis, other sectors, such as the construction industry suffered, too, with the state of the financial system also contributing in this direction.

Higher risk and uncertainty for the future of the sector would imply that organisations that supply, or are being supplied by, the focal manufacturers are also at risk (Bozarth and Berry, 1997). It is in response to this risk that we hypothesise manufacturers would pursue a policy of vertical integration to increase resilience. The rationale behind such a decision is *asset specificity*, which is defined as 'durable investments that are undertaken in support of particular transactions' (Williamson, 1985), which are executed frequently and have highly uncertain outcomes. These transactions are performed more efficiently within the firm (Gulbrandsen et al., 2009).

The effect of a crisis on a firm is also likely to be contingent upon the institutional model of the country. In Greece, there has been an asymmetry in the size of companies in the value chain of most industries (Makridakis et al., 1997). Distributors have been smaller and more localised compared to manufacturers. Small companies, which were closer to the end user, proved to be less resilient to the crisis due to their weak financial position and their dependence on credit (European Commission, 2013). Changes in the tax system and stricter monitoring also had a negative effect on the ability of smaller firms to survive. As a result, we argue that manufacturers would increase their resilience by taking over activities, which would normally be undertaken by their customers, that is, they would forward integrate, for two reasons. First would be to reduce market-related risk and increase their resilience by ensuring continuity of distribution of their products. Acquiring smaller organisations would reduce the risk associated with the increased vulnerability due to size. Second would be to gain additional value from these activities to compensate for any reduced sales, that is, forward integration would allow them to absorb some of the gains of their customers and distributors and hence increase their ability to respond to the crisis.

Based on these arguments, the following hypothesis is made:

Hypothesis 1: Organisations that realise they belong to a sector that is negatively affected by an economic crisis are more likely to come closer to the end user, that is, they are more likely to forward vertically integrate.

6.4.2 COMPLEXITY, RESILIENCE AND INTEGRATION

In general, if significant in-house work is required, forward integration results in reduced ability to increase product variety (Rothaermel, 2012). In-house product development, marketing and distribution of a variety of products require investments in specialised assets, as well as development of the appropriate culture and cognitive processes (Volberda, 1998; Sanchez and Heene, 2002). Consequently, companies that already have these capabilities and mind frames can adopt strategies of forward integration without additional burden, especially in times of economic crises where the range and size of demand is limited (Chong et al., 2014).

Managers in companies associated with a variety of products and markets develop cognitive abilities in dealing with complex issues. The complexity of the issues challenges the ability of managers to make sense of diverse environmental signals required for producing diverse strategies (strategic flexibility). The *complexity* and *centrality* (focus and specialisation) of mental models (Nadkarni and Narayanan, 2007) depends on the managers' past experience with environment-induced strategic issues, including the management of a diverse product range (Carley and Palmquist, 1992; Reger and Palmer, 1997). Consequently, operations functions that have to deal with a wide range of products develop complex cognitive schemas that allow them to respond to novel situations efficiently and effectively (McCann and Selsky, 2012), making the entire organisation more resilient (Weick and Sutcliffe, 2007).

Thus, our second hypothesis is:

Hypothesis 2: Organisations that prior to the crisis built resilience by focusing on developing systems for delivering a high variety of products are more likely to come closer to the end user, that is, they are more likely to forward vertically integrate.

6.4.3 CAPABILITY, PATH DEPENDENCE AND INTEGRATION

Continuing the above argumentation, path dependence exists not only in the range and pattern of decisions and the process of managerial decision making, but also, inevitably, in the results/outcomes of decisions and their implementation. The majority of strategy decisions concern investments in resources and capabilities. Companies investing in specific sets of resources and capabilities (competencies) find it difficult to redirect their investments towards significantly different ones. The relatedness between a firm's competence and the competence required for undertaking a forward or backward integration activity is related to the difficulty in undertaking the new activity and logically to the decision to integrate (Winter, 1988; Gulbrandsen et al., 2009). More specifically, manufacturing companies that have concentrated on their production activity and invested accordingly in associated methods and technologies (exploitation of production-oriented strategies) to build resilience through reliability and robustness of a core activity will face difficulty in adopting forward integration strategies that require, in addition to complex cognitive schemas, different tangible and intangible resources and related capabilities. In addition to being robust, strategies should be adaptive through the cultivation of a diverse range of possibilities (Beinhocker, 1999) in order to augment their resilience (Reinmoeller and van Baardwijk, 2005).

Hence, in the context of prior and after the crisis strategies, we can hypothesise:

Hypothesis 3: Organisations that prior to the crisis focused on the robustness facet of resilience through the development of production capabilities are less likely to respond to the crisis by coming closer to the end user, that is, they are less likely to forward vertically integrate.

6.4.4 TECHNOLOGY AND FORWARD INTEGRATION

Our last hypothesised relationship explores the impact of the development of technologies for improving the management of processes to the pursuit of a policy of vertical integration. In general, investment in new process technology can improve the efficiency of business (Hayes et al., 2005). In addition, technology could support the development of unique capabilities (asset specificity) for responding faster and more efficiently to customer needs (Coates and McDermott, 2002).

However, when these investments are made in response to a crisis, it would be reasonable to expect that processes and the associated technology will be relatively at a more advanced level when compared with other companies within the same supply chain. On the customer side, which is the focus of the chapter, this would suggest that manufacturers that improve their process technologies will be better equipped to both understand and respond to changing needs of the end user, a premise that has been supported in the strategy literature (Dosi et al., 1992; Frohlich and

Westbrook, 2001; Devaraj et al., 2007). Put differently, manufacturers, who invest in new technologies for process improvement after a crisis in anticipation of changes in the downstream part of the value chain, will have an increased ability to understand their customers' customers. This increased ability would create an incentive to either bypass their intermediate customer or acquire their activities, as strategic choice or out of necessity, that is, it will encourage the implementation of a policy of forward vertical integration. Broken business links in the supply chain that were based on personal long-term relationships are difficult to repair by equally trusting new partners. Hence, it is logical to rely on the impersonality of information technology for the substitution of failed links.

A reason that could increase the likelihood of the implementation of a policy of forward vertical integration relates to the reduction of supply-chain risk by objectifying transactions through technology (Kallinikos, 2011). Manufacturers who have been able to respond to the crisis by investing in process technologies that are in short 'distance' of their existing resources and capabilities, for example, implementation of ERP technologies (Slack and Lewis, 2002), are likely to be in stronger financial positions relative to their supply-chain partners. When such partners are at risk of reduced performance, or even failure, it would be reasonable to expect that the focal manufacturers would be more incentivised to acquire control of their customers' activities (Enkel et al., 2005) and integrate them through technology. The motivation to reduce risk and increase resilience would therefore increase the incentive for pursuing a policy of integrating in a way that brings the focal manufacturer closer to the end customer, that is, to forward vertically integrate. In addition, given the formalised relationship among employees within Greek firms, it seems easier to codify the information flow in the supply chain. From the above arguments we deduct our fourth hypothesis:

> Hypothesis 4: Organisations that after the crisis focus on developing new technologies for improving process management are also likely to respond to the crisis by coming closer to the end user, that is, they are more likely to forward vertically integrate.

6.5 METHODOLOGY

The aim of this chapter is to explore which dimensions of built resilience enable firms to adopt strategies of vertical integration for surviving the economic crisis. To meet this aim and to ensure the generalizability of the findings, a questionnaire survey was conducted. Questionnaire surveys have been employed by several studies in the past and are increasingly seen as an established research technique for operations, and technology management studies (e.g. Frohlich and Westbrook, 2001; Da Silveira and Cagliano, 2006; Leenders et al., 2007; Lages et al., 2008; Tsinopoulos and Al Zu'bi, 2012).

6.5.1 RESEARCH CONTEXT

Data was gathered via a self-administered questionnaire sent to Greek manufacturing companies. In particular, questionnaires were sent to the 150 largest manufacturers in Greece in terms of revenues in 2010. The survey was sent to previously identified operations managers of each company. After follow-up phone calls and a second round of mail shot, 86 usable questionnaires were received. Manufacturers of both consumer and capital goods were used to test the four hypotheses stated above. The survey was carried out in the period between December 2011 and May 2012, which was sufficiently within the period of the financial crisis in Greece.

Of the responding companies, 39% employed 250–300 people, whereas 30% employed between 500 and 1000. Only 11 companies employed more than 1000. The majority of the responding companies (43%) had revenues between 100 and 500 million euros in 2010, while only 12 companies (14%) had revenues exceeding 500 million euros. Food and beverages and metal processing and equipment were the two sectors with the highest representation in the sample (33% and 29%, respectively).

6.5.2 MEASURES

The questionnaire sent was multi-faceted, consisting of six sections. Likert-type scales of range 1–5 were used. The first section asked for general information about the company (size, industry, etc.), the second about the perceived changes in the institutional environment after the crisis (relations with competitors, financial institutions, customers/demand, etc.), the third about the operations strategy priorities (cost, flexibility, speed, dependability and quality) and the corresponding priorities of capabilities development (capacity management, technology management, supply-chain management and HR/organisational development) before the crisis. The fourth section repeated the same questions for what holds after the crisis. Two additional sections asked questions related to the practice of operations management, which were not, however, directly related to the objective of this study.

There were four independent and one dependent variables of interest to the study. The first related to the variety of products and was measured by asking operations managers to rate the degree to which their systems were sufficiently flexible prior to the crisis. The second related to the development of production capabilities prior to the crisis. This was measured by asking operations managers to rate the degree to which they had invested for developing such capabilities prior to the onset of the financial crisis. The third variable related to the development of the implementation and development of new technologies for improving processes. Again this was measured by asking respondents to rate the degree to which they had implemented new technologies for improving processes prior to the crisis. The final independent variable related to the sector in which the organisation belonged. As with the other independent variables this was measured by asking respondents to rate the degree to which they felt that their sector has been affected by the crisis.

The dependent variable related to the degree to which the organisation had followed a policy of vertical integration after the crisis. We also included the impact of the crisis as a control variable. The impact of the crisis may have had a disproportionate impact on some sectors.

6.6 FINDINGS

We start by providing a description of the general trends identified through the survey. To do so we use two diamond diagrams, Figures 6.1 and 6.2, which depict the trends before and after the crisis. A total of 83.1% of respondents indicated a general trend in reduced profitability and 87.9% problems with the availability of cash. 89.1% indicated intensification of competition and 91.6% a rapid drop in demand. In general, the answers to the questionnaire indicated an increased importance given to the operations strategy priorities after the crisis; towards cost reduction and increased flexibility (Figure 6.1). However, as far as the means to achieve operations strategy objectives were concerned, there was a clear shift of priorities towards supply-chain management, HR and organisational issues at the expense of capacity and process technology management (Figure 6.2).

Descriptive statistics and correlations of the answers are shown in Table 6.2.

Linear regression analysis provided support for our hypotheses, which predict the situational factors that can increase the probability of an organisation to respond to a crisis by vertical integration. The results are shown in Table 6.3.

The first hypothesis predicted that when managers perceive that their organisations belong in a sector that has been negatively affected by an economic crisis, they are more likely to respond by coming closer to the end user, that is, they are more likely to forward vertically integrate. As shown in Table 6.3, the coefficient is positive (0.355) and statistically significant.

Hypothesis 2 predicted that organisations that prior to the crisis built resilience by focusing on developing systems for delivering a high variety of products are more likely to respond to the crisis by coming closer to the end user, that is, they are more likely to forward vertically integrate. As shown in Table 6.3, the coefficient is positive (0.286) and significant at the 5% level providing support for this hypothesis.

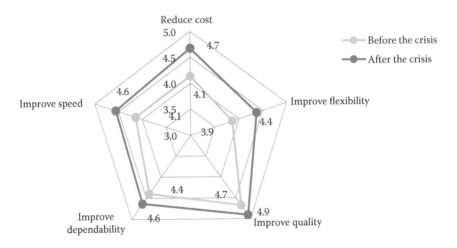

FIGURE 6.1 Operations strategic priorities before and after the crisis.

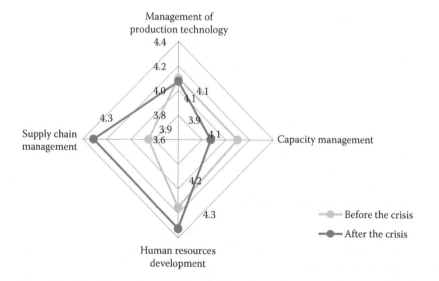

FIGURE 6.2   Effect of crisis on operations strategy decision areas.

Hypothesis 3 predicted that organisations which prior to the crisis focused on the robustness facet of resilience by developing production capabilities are less likely to respond by coming closer to the end user, that is, they are less likely to forward vertically integrate. As shown in Table 6.3 the coefficient is negative (-0.225) and statistically significant providing support for the third hypothesis.

The fourth hypothesis predicted that organisations which after the crisis focused on developing new technologies for improving process management are also likely to have responded to the crisis by coming closer to the end user, that is, they are more likely to forward vertically integrate. The model indicates that the relative coefficient is positive (0.296) and statistically significant ($p < 0.05$).

The above findings indicate that the likelihood of the surveyed organisations to respond to the crises by extending their operations, forward or backward, depends on a combination of operations strategic choices related to resilience capabilities made both prior to the onset of and during the unfolding of the crisis. These results also indicate that these choices will differ depending on the severity of the impact of the crisis on the sector to which organisations belong.

6.7   CONCLUSIONS

In this chapter, we focused on the impact economic crises have on a firm's environment and its strategic management and how the latter is modulated by the existence of resilience-related capabilities. To explore which dimensions of built resilience enable firms to adopt strategies of vertical integration for surviving economic crises we conducted a survey of Greek manufacturing companies. We identified two pre-crisis strategies that aimed at building resilience. The first relates to developing

TABLE 6.2
Descriptive Statistics and Correlations

Variable	Mean	Std	IC	FF	PMF	PTI	SNA	FVI
Increased competition (IC)	3.7442	1.149	1					
Flexibility focus (FF)	3.861	0.959	-0.097	1				
Production management focus (PMF)	4.256	0.722	-0.147	-0.175	1			
Process technology investment (PTI)	4.30	0.798	-0.069	-0.067	0.313[a]	1		
Degree to which sector has been negatively affected (SNA)	3.721	1.325	0.617[b]	-0.012	0.026	-0.075	1	
Forward vertical integration response (FVI)	2.605	1.391	-0.005	0.240[a]	-0.039	0.194	0.195	1

[a] Correlation significant at the 5% level.
[b] Correlation significant at the 1% level.

TABLE 6.3
Regression Analysis, Vertical Integration Response

Variable	Standardised Coefficient	Standard Error	Significance
Constant		1.260	0.811
Increased competition	−0.209	0.157	0.111
Flexibility focus	0.286[a]	0.149	0.007
Production management focus	−0.225[a]	0.211	0.044
Process technology investment	0.296[a]	0.186	0.007
Degree to which sector has been negatively affected	0.355[a]	0.135	0.007
R^2	0.205		
Model significance	0.002		
F	4.125		

[a] Significant at the 5% level.

systems for delivering a variety of products and the second relates to the development of production capabilities. We argued that both helped explicitly or tacitly to build capabilities to withstand the shocks of economic crises and continue to operate effectively.

Our results indicate that companies' strategic response to the crisis will vary depending on their selected pre-crisis resilience strategy. First, when they have invested in systems for delivering a high variety of products, they are likely to adopt strategies of forward integration. Second, when they have invested in the development of production systems, they are less likely to forward integrate. It is the specific effects that crises have on specific dimensions of these institutional systems that make companies adopt these strategies. In addition, our empirical data indicate that companies operating in sectors more severely hit by the crisis are more likely to adopt such strategies and that technology plays an important role in the deployment of forward integration in the specific context.

Our study contributes to the literature in two ways. First, we have explored the impact of resilience on an organisation's strategic response following an economic crisis. Both our theoretical development and our results indicate that the ability of an organisation to respond to such crises is dependent on strategic decisions which increase an organisation's ability to produce a variety of new products and on the development of production capabilities. Second, we have explained how strategies aimed at building resilience, such as increased flexibility and investment, will also determine the strategic response following the crisis. Our results therefore indicate that there is a path dependence on the response of individual companies, a phenomenon that is reinforced by the technology choices organisations take during the immediate aftermath of the crisis.

Given the above conclusions we would make the following recommendations to operations managers in relation to both their efforts to build resilience and their immediate reactions in the aftermath of a crisis. The first relates to the rationale and motivation of building resilience in the first place. It is clear from our results that any choices made before a crisis will require a certain strategic response after the crisis occurs. We would therefore recommend managers to explore their long-term willingness to vertically integrate or not. When there is a desire to do so, then the development of a system that increases the variety of products would be beneficial.

The second recommendation relates to the potential responses following a crisis. Given the patterns we identified in this study, we would encourage managers to prepare plans which make more effective use of the responses we have identified here. If, for instance, a system is in place which increases the variety of products, then managers could prepare their organisations for opportunities of vertical integration in the aftermath of the crisis. Such approaches would help to both build more appropriate resilience strategies and develop appropriate strategies post the crisis.

6.8   LIMITATIONS AND FUTURE RESEARCH

Inevitably, our study has some limitations, within the context of which our results need to be interpreted. The first relates to the context of the research. The study was conducted using data from Greek companies only. Although the timing of the study provided a great opportunity to explore the interplay between the building of resilience strategies, such crises can be context specific. As a result, some of our results may not automatically be generalizable across other countries. We would therefore encourage researchers to replicate aspects of our study in other countries as well.

A second limitation results from the nature of the data. We asked respondents to evaluate aspects of their processes both before and after the crisis. Although this approach provided us with some rich data, we cannot claim that we have eliminated the potential for the usual limitations associated with this kind of data collection process, for example, common method bias. We would therefore encourage research which uses both objective and longitudinal data.

One additional avenue for future research that emanates from our study relates to the strategies used to build resilience prior to the crisis. Flexibility, investment in new technologies and the development of production facilities would inevitably increase an organisation's ability to respond effectively to a crisis. However, these are not the only strategies. We would therefore encourage future studies to explore some additional strategies, particularly those which relate more broadly to the supply chain. Given that such crises are likely to have an impact across the supply chain, building resilience which spans organisational boundaries would be an inevitable direction to explore.

ACKNOWLEDGEMENTS

We would like to thank E. Peraki and M. Kapnisi for their assistance during the research.

REFERENCES

Ackerman, S.R. (Ed.) 2006, *International Handbook on the Economics of Corruption*, Yale University Press, New Haven.

Adamides, E.D., Stamboulis, Y. and Kanellopoulos, V. 2003, Economic integration and strategic change: The role of managers' mental models, *Strategic Change*, Vol. 12, No. 2, pp. 69–82.

Aghion, P. and Roulet, A. 2011, *Repenser l'État. Pour une social-démocratie de l'innovation*, Éditions du Seuil, Paris.

Almond, P. 2011, Nations, regions and international HRM, in: Edwards, T. and Rees, C. (Eds.), *International Human Resource Management*, FT/Prentice Hall, New York, pp. 50–66.

Amable, B. 2003, *The Diversity of Modern Capitalism*, Oxford University Press, Oxford.

Arya, A. and Mittendorf, B. 2013, The changing face of distribution channels: Partial forward integration and strategic inventory, *Production and Operations Management*, Vol. 29, No. 5, pp. 1077–1088.

Beinhocker, E.D. 1999, Robust adaptive strategies, *Sloan Management Review*, Vol. 40, No. 3, pp. 95–106.

Bhamra, R., Dani, S. and Burnard, K. 2011, Resilience: The concept, a literature review and future directions, *International Journal of Production Research*, Vol. 49, No. 18, pp. 5375–5393.

Bozarth, C.C. and Berry, W.L. 1997, Measuring the congruence between market requirements and manufacturing: A methodology and illustration, *Decision Sciences*, Vol. 28, No. 1, pp. 121–150.

Brahm, F. and Tarziján, J. 2014, Transactional hazards, institutional change, and capabilities: Integrating the theories of the firm, *Strategic Management Journal*, Vol. 35, No. 2, pp. 224–245.

Burgelman, R.A. 2005, The role of strategy making in organizational evolution, in: Bower, J.L. and Gilbert, C.G. (Eds.), *From Resource Allocation to Strategy*, Oxford University Press, Oxford, pp. 38–70.

Cachon, G.P. 2003, Supply chain coordination with contracts, in: Graves, S. and de Kok, T. (Eds.), *Handbooks in Operations Research and Management Science: Supply Chain Management*, North-Holland, Amsterdam, pp. 227–339.

Cadeaux, J. and Ng, A. 2012, Environmental uncertainty and forward integration in marketing: Theory and meta-analysis, *European Journal of Marketing*, Vol. 46, No. 1/2, pp. 5–30.

Carley, K. and Palmquist, M. 1992, Extracting, representing, and analyzing mental models, *Social Forces*, Vol. 70, No. 3, pp. 601–636.

Chang, S.-J. 2003, *Financial Crisis and Transformation of the Korean Business Groups: The Rise and Fall of Chaebols*, Cambridge University Press, Cambridge.

Chong, Y.Q., Wang, B., Tan, G.L.Y. and Cheong, S.A. 2014, Diversified firms on dynamical supply chain cope with financial crisis better, *International Journal of Production Economics*, Vol. 150, pp. 239–245.

Christopher, M. and Lee, H.L. 2004, Mitigating supply chain risk through improved confidence, *International Journal of Physical Distribution and Logistics Management*, Vol. 34, No. 5, pp. 388–396.

Coates, T.T. and McDermott, C.M. 2002, An exploratory analysis of new competencies: A resource based view perspective. *Journal of Operations Management*, Vol. 20, No. 5, pp. 435–450.

Conway, P., de Rosa, D., Nicoletti, G. and Steiner, F. 2006, Regulation, competition and productivity convergence, *OECD Economics Department Working Paper*, No. 509.

Coutu, D. 2002, How resilience works, *Harvard Business Review*, Vol. 80, No. 5, pp. 46–50.

Da Silveira, G.J.D. and Cagliano, R. 2006, The relationship between interorganizational information systems and operations performance, *International Journal of Operations and Production Management*, Vol. 26, No. 3, pp. 232–281.

David, R.J. and Han, S.-K. 2004, A systematic assessment of the empirical support for transaction cost economics, *Strategic Management Journal*, Vol. 25, pp. 39–58.

Devaraj, S., Krajewski, L. and Wei, J.C. 2007, Impact of eBusiness technologies on operational performance: The role of production information integration in the supply chain, *Journal of Operation Management*, Vol. 25, No. 6, pp. 1199–1216.

Diez-Val, K. 2007, Explaining vertical integration strategies: Market power, transactional attributes and capabilities, *Journal of Management Studies*, Vol. 44, No. 6, pp. 1017–1040.

DiMaggio, P. and Powell, W. 1983, The iron cage revisited: Institutionalized isomorphism and collective rationality in organisational fields, *American Sociological Review*, Vol. 48, pp. 147–160.

Ding, L. and Mahbubani, J. 2013, The two-stage decision model of vertical integration, *Management Decision*, Vol. 51, No. 2, pp. 306–320.

Dosi, G., Teece, D.J. and Winter, S. 1992, Toward a theory of corporate coherence: Preliminary remarks, in: Dosi, G., Gianetti, R. and Toninelli, P.A. (Eds.), *Technology and Enterprise in an Historical Perspective*, Clarendon, Oxford, pp. 185–211.

Enkel, E., Kauch, C. and Gassman, O. 2005, Managing the risk of customer integration, *European Management Journal*, Vol. 23, No. 2, pp. 203–213.

European Commission 2013, SBA Fact Sheet 2013 – Greece, available at http://ec.europa.eu/enterprise/policies/sme/facts-figures-analysis/performance-review/files/countries-sheets/2013/greece_en.pdf (accessed 7-8-2014).

Flynn, B., Huo, B. and Zhao, X. 2010, The impact of supply chain integration on performance: A contingency and configuration approach, *Journal of Operations Management*, Vol. 28, No. 1, pp. 58–71.

Frohlich, M.T. and Westbrook, R. 2001, Arcs of integration: An integration study of supply chain strategies, *Journal of Operations Management*, Vol. 19, No. 2, pp. 185–200.

Gibson, C. A. and Tarrant, M. 2010, A conceptual models' approach to organisation resilience. *Australian Journal of Emergency Management*, Vol. 25, No. 2, pp. 6–12.

Grossler, A., Grübner, A. and Milling, P.M. 2006, Organisational adaptation processes to external complexity, *International Journal of Operations and Production Management*, Vol. 26, No. 3, pp. 254–281.

Grossman, S.J. and Hart, O.D. 1986, The costs and benefits of ownership: A theory of vertical and lateral integration, *Journal of Political Economy*, Vol. 94, No. 4, pp. 691–719.

Guan, W. and Rehme, J. 2012, Vertical integration in supply chains: Driving forces and consequences for a manufacturer's downstream integration, *Supply Chain Management: An International Journal*, Vol. 17, No. 2, pp. 187–201.

Gulbrandsen, B., Sandvik, K. and Haugland, S.A. 2009, Antecedents of vertical integration: Transaction cost economics and resource-based explanations, *Journal of Purchasing and Supply Chain Management*, Vol. 15, pp. 89–102.

Hall, P.A. and Soskice, D. 2001, An introduction to varieties of capitalism, in: Hall, P.A. and Soskice, D. (Eds.), *Varieties of Capitalism: The Institutional Foundations of Comparative Advantage*, Oxford University Press, Oxford, pp. 1–68.

Hamel, G. and Välikangas, L. 2003, The quest for resilience, *Harvard Business Review*, Vol. 81, No. 9, pp. 52–63.

Hayes, R.H. and Wheelwright, S.C. 1984, *Restoring our Competitive Edge: Competing through Manufacturing*, Wiley, New York.

Hayes, R., Pisano, G., Upton D. and Wheelwright S. (2005), *Operations, Strategy, and Technology: Pursuing the Competitive Edge*, Wiley, New York.

Heide, J.B. and Stump, R.L. 1995, Performance implications of buyer-supplier relationships in industrial markets: A transaction cost explanation, *Journal of Business Research*, Vol. 32, pp. 57–66.

Hendricks, K.B., Singhal, V.R. and Zhang, R. 2009, The effect of operational slack, diversification and vertical relatedness on the stock market reaction to supply chain disruptions, *Journal of Operations Management*, Vol. 27, No. 3, pp. 233–246.

Jacobides, M. and Winter, S. 2005, The co-evolution of capabilities and transaction costs: Explaining the institutional structure of production, *Strategic Management Journal*, Vol. 26, No. 5, pp. 395–413.

Kallinikos, J. 2011, *Governing through Technology: Information Artefacts and Social Practice*, Palgrave Macmillan, Basingstoke, UK.

Kauppi, K. 2013, Extending the use of institutional theory in operations and supply chain management: Review and research suggestions, *International Journal of Operations and Production Management*, Vol. 33, No. 10, pp. 1318–1345.

Kaya, M. and Özer, Ö. 2012, Pricing in business-to-business contracts: Sharing risk, profit and information, in: Özer, Ö. and Phillips, R. (Eds.), *The Oxford Handbook of Pricing Management*, Oxford University Press, Oxford, pp. 738–783.

Koufteros, X.A., Cheng, T.C.E. and Lai, K.H. 2007, Black-box and grey box supplier integration in product development: Antecedents, consequences and the moderating role of the firm size, *Journal of Operations Management*, Vol. 25, No. 4, pp. 847–870.

Lages, L.F., Jap, S.D. and Griffith, D.A. 2008, The role of past performance in export ventures: A short-term reactive approach, *Journal of International Business Studies*, Vol. 39, No. 2, pp. 304–325.

Leenders, R.T., van Engelen, J.M.L. and Kratzer, J. 2007, Systematic design methods and the creative performance of new product teams: Do they contradict or complement each other? *Journal of Product Innovation Management*, Vol. 24, No. 2, pp. 166–179.

Lehrer, M. 2001, Macro-strategies of capitalims and micro-varieties of strategic managemnt in European airlines, in: Hall, P.A. and Soskice, D. (Eds.), *Varieties of Capitalism: The Institutional Foundations of Comparative Advantage*, Oxford University Press, Oxford, pp. 361–386.

Lin, Y-T., Parlaktürk, A.K. and Swaminahanan, J.M. 2014, Vertical integration under competition: Forward, backward, or no integration? *Production and Operations Management*, Vol. 23, No. 1, pp. 19–35.

Makridakis, S., Caloghirou, Y., Papagiannakis, L. and Trivellas, P. 1997, The dualism of Greek firms and management: Present state and future implications, *European Management Journal*, Vol. 15, No. 4, pp. 381–402.

McCann, J. and Selsky, J.W. 2012, *Mastering Turbulence: The Essential Capabilities of Agile and Resilient Individuals, Teams and Organizations*, Jossey-Bass, San Francisco, CA.

Milgrom, P. and Roberts, J. 1992, *Economics, Organization and Management*, Prentice Hall, Englewood Cliffs, NJ.

Mintzberg, H. 1994. *The Rise and Fall of Strategic Planning*, The Free Press, New York, NY.

Mitroff, I.I., Pauchant, T.C., Finney, M. and Pearson, C. 1989, Do (some) organisations cause their own crises? The cultural profiles of crisis-prone *vs* crisis-prepared organisations, *Industrial Crisis Quarterly*, Vol. 3, No. 4, pp. 269–283.

Nadkarni, S. and Narayanan, V. K. 2007, Strategic schemas, strategic flexibility, and firm performance: The moderating role of industry clockspeed, *Strategic Management Journal*, Vol. 28, No. 3, pp. 243–270.

Nagurney, A., Cruz, J., Dong, J. and Zhang, D. 2005, Supply chain networks, electronic commerce, and supply side and demand side risk. *European Journal of Operational Research*, Vol. 164, No. 1, pp. 120–142.

North, D.C. 1990, *Institutions, Institutional Change and Economic Performance*, Cambridge University Press, Cambridge.

Osegowitsch, T. and Madhok, A. 2003, Vertical integration is dead, or is it? *Business Horizons*, Vol. 46, No.2, pp. 25–34.

Papadakis, V.M. 2006, Do CEOs shape the process of making strategic decisions? Evidence from Greece, *Management Decision*, Vol. 44, No. 3, pp. 367–394.

Psychogios, A.G. and Szamosi, L.T. 2007, Exploring the Greek national business system, *EuroMed Journal of Business*, Vol. 2, No. 1, pp. 7–22.

Rangan, V.K., Corey, E.R. and Cespedes, F. 1993, Transaction cost theory: Inferences from clinical field research on downstream integration, *Organization Science*, Vol. 4, No. 3, pp. 454–477.

Reger, R.K. and Palmer, T.B. 1997, Managerial categorization of competitors: Using old maps to navigate new environments, *Organization Science*, Vol. 7, pp. 22–39.

Reinmoeller, P. and van Baardwijk, N. 2005, The link between diversity and resilience, *Sloan Management Review*, Vol. 46, No. 4, pp. 61–65.

Rhee, M. and Mehra, S. 2006, Aligning operations, marketing and competitive strategies to enhance performance: An empirical test in the retail banking industry, *Omega – The International Journal of Management Science*, Vol. 34, No. 5, pp. 505–515.

Richardson, J. 1996, Vertical integration and rapid response in fashion apparel, *Organization Science*, Vol. 7, No. 4, pp. 400–412.

Rothaermel, F.T. 2012, *Strategic Management: Concepts*, Irwin/McGraw-Hill, New York.

Sanchez, R. 1995, Strategic flexibility in product competition, *Strategic Management Journal*, Vol. 16, S1 (Special Issue), pp. 135–159.

Sanchez, R. and Heene, A. 2002, Managing strategic change: A systems view of strategic organizational change and strategic flexibility, in: Morecroft, J., Sanchez, R. and Heene, A. (Eds.), *Systems Perspectives on Resources, Capabilities, and Management Processes*, Pergamon, Oxford.

Seville, E., Brunsdon, D., Dantas, A., Le Masurier, J., Wilkinson, S. and Vargo, J. 2008, Organisational resilience: Researching the reality of New Zealand organisations, *Journal of Business Continuity and Emergency Management*, Vol. 2, No. 2, pp. 258–266.

Slack, N. and Lewis, M. 2002, *Operations Strategy*, FT- Prentice Hall, Harlow, UK.

Schmiermann, M. 2008, Enterprises by size class, *Industry, Trade and Services – Statistics in Focus* 8, Eurostat, Luxembourg.

Soskice, D. 2009, Varieties of capitalism; varieties of reform, in: Hemerjick, A., Knapen, B. and Van Doorne, E. (Eds.), *Aftershocks: Economic Crisis and Institutional Choice*, Amsterdam University Press, Amsterdam, pp. 133–142.

Spanos, Y., Prastacos, G. and Papadakis V. 2001, Greek firms and EMU: Contrasting SME and large-sized enterprises, *European Management Journal*, Vol. 19, No. 6, pp. 638–648.

Sutcliffe, K.M. and Zaheer, A. 1998, Uncertainty in the transaction environment: An empirical test, *Strategic Management Journal*, Vol. 19, pp. 1–23.

Swamidass, P.M. and Newell, W.T. 1987, Manufacturing strategy, environmental uncertainty and performance: A path analytic model, *Management Science*, Vol. 33, No. 4, pp. 509–524.

Teece, D., Pisano, G. and Shuen, A. 1997, Dynamic capabilities and strategic management, *Strategic Management Journal*, Vol. 18, No. 7, pp. 509–533.

Thomson, A.A. and Strickland, A. J. 2001, *Strategic Management: Concepts and Cases* (13th Edition), Irwin/McGraw-Hill, Chicago.

Toulan, O.N. 2002, The impact of market liberalization on vertical scope: The case of Argentina, *Strategic Management Journal*, Vol. 23, pp. 551–560.

Tsinopoulos, C. and Al Zu'bi, Z.b.M.F. 2012. Clockspeed effectiveness of lead users and product experts, *International Journal of Operations and Production Management*, Vol. 32, No. 9, pp. 1097–1118.

Tylecote, A. and Visintin, F. 2008, *Corporate Governance, Finance and the Technological Advantage of Nations*, Routledge, London.

Välikangas, L. and Romme, G.L. 2012, Building resilience capabilities at 'Big Brown Box, Inc.', *Strategy & Leadership*, Vol. 40, No. 4, pp. 43–45.

Volberda, H.W. 1998, *Building the Flexible Firm: How to Remain Competitive*, Oxford University Press, Oxford.

Volberda, H.W. 2003, Strategic flexibility: Creating dynamic competitive advantages, in: Faulker, D.O. and Campbell, A. (Eds.), *The Oxford Handbook of Strategy, Volume II: Corporate Strategy*, Oxford University Press, Oxford, pp. 447–506.

Wagner, S.M. and Bode, C. 2006, An empirical investigation into supply-chain vulnerability, *Journal of Purchasing & Supply Management*, Vol. 12, No. 6, pp. 301–312.

Walker, G. and Weber, D. 1984, A transaction cost approach to make versus buy decision, *Administration Science Quarterly*, Vol. 29, pp. 589–596.

Weick, K.E. and Sutcliffe, K.M. 2007, *Managing the Unexpected: Resilient Performance in an Age of Uncertainty* (2nd Edition), Jossey-Bass, San Francisco, CA.

Williamson, O.E. 1985, *The Economic Institutions of Capitalism*, Free Press, New York.

Williamson, O.E. 1990, A comparison of alternative approaches to economic organization, *Journal of Institutional and Theoretical Economics*, Vol. 146, pp. 61–71.

Winter, S.G. 1988, On Coase, competence, and the corporation. *Journal of Law, Economics, and Organization*, Vol. 4, No. 1, pp. 163–180.

Wise, R. and Baumgartner, P. 1999, Go downstream: The new profit imperative in manufacturing, *Harvard Business Review*, Vol. 77, No. 5, pp. 133–141.

Witcher, B.J. and Chau, V.S. 2012, Varieties of capitalism and strategic management: Managing performance in multinationals after the global financial crisis, *British Journal of Management*, Vol. 23, pp. S58–S73.

Woods, D. D. and Wreathall, J. 2008, Stress-strain plots as a basis for assessing system resilience, in: Hollnagel, E., Nemeth, C.P. and Dekker, S. (Eds.), *Remaining Sensitive to the Possibility of Failure*, Ashgate, Padstow, pp. 143–158.

Workers-participation.eu 2014, *Board-Level Representation* (Workers-participation.eu is an information service of the European Trade Union Institute), available at http://www.worker-participation.eu/National-Industrial-Relations/Across-Europe/Board-level-Representation2 (accessed 17-7-2014).

7 Resilient Requirements for Emergency First Responders

Lili Yang

CONTENTS

7.1   INTRODUCTION

7.1.1   CONTEXT

Disasters such as fires, floods, earthquakes, civil war or terrorist attacks may cause crisis situations. Regardless of the origin, crisis situations are often accompanied by uncertainty of how the disaster will develop, a rapid pace of response operations and the possibility of serious loss of human lives and property if not responded to properly. For these reasons the ability and efficiency of responding to crises and unexpected events has become increasingly important throughout the world, particularly in the United States and the United Kingdom in response to the 09/11 terrorist attack on the World Trade Centre and the 7 July London bombings. These events are etched in memory and were those in which humans were ill equipped to respond. Good situational awareness and decision-making support are important factors for minimising property damage and injury, and for saving people's lives. To provide adequate situational awareness and decision-making support to manage crisis situations, researchers and practitioners in disaster management have urged attention to emergency information requirement capturing, information presentation, information sharing and consequently information-driven decision support through the research and development of emergency response information systems (ERISs). ERISs should support first responders by enhancing their situational awareness which will lead to better decision making (Klann, 2008).

7.1.2   DYNAMIC EMERGENCY INFORMATION

Different from the ordinary information for office use, emergency information may be utilised in an extreme and stress-filled environment, including not only static information such as road maps and building floor plans, but also dynamic and real-time information such as information about the latest disaster developments and the current locations of emergency personnel and resources. As an emergency evolves, requirements (both informational and logistical) may change resulting in necessary modifications of the response workflow (Wang et al., 2008, 2009). An investigation of first responders' requirements in a Dutch emergency response (ER) case illustrated that much of the information first responders request during a crisis can be considered dynamic information and is needed almost instantaneously (Diehl et al., 2006; Yang et al., 2013).

7.1.3   REQUIREMENTS OF INFORMATION SHARING

Furthermore, the scale and demands of disasters require the participation of many different response organisations. Depending upon the severity of the crisis, ER may involve numerous organisations including multiple layers of government, public authorities such as fire and rescue services (FRSs) and police forces, commercial entities, volunteer organisations such as Red Cross, media organisations and the public. These entities work together to save lives, preserve infrastructure and community resources and to re-establish normality within the population. The efficacy of response is mainly determined by the ability of decision makers to understand the crisis at hand and the state of the available resources to make vital decisions (Mehrotra et al., 2004). The quality of these decisions in turn depends upon the timeliness, amount and accuracy of the information available to the responders. Making good decisions really relies on the available information and the ability of decision makers to cope with the demands imposed in the management of ERs (Danielsson, 1998). Therefore, information sharing among different response organisations is essential to success of the ERs. Information sharing among them cannot occur overnight and must be in place before a disaster occurs, be able to be easily used during the disaster and be maintained after the disaster (Yang, 2007).

The objectives of information sharing in ERs focus on supporting incident commanders in decision making, guiding and protecting emergency responders in response operations and protecting members of the public who may be located in or near in the disaster scene.

7.2 ER IN THE UNITED KINGDOM

Like most of the countries in the world, the FRSs in the United Kingdom have to follow a set of strict work procedures in their ER operations, from handling an emergency call, to dispatching ER forces, to on-site preliminary situation assessment and then to crisis response (Fire Service Manual, 2008). These procedures are extensively explained in all kinds of documents, some of which are available for access on the Internet. This section does not aim to provide a further illustration of the ER work practice, which is available from many sources (Berrouard et al., 2006; Diehl et al., 2006; Landgren, 2006), but identifies the change of priorities of first responders during the different stages of their operations.

ER operations are triggered by 999 calls handled by a command centre of the FRS in the United Kingdom. The command centre has the ability to dispatch police cars, ambulances and fire engines. A certain number of fire engines from the nearest and available fire brigades are dispatched to the incident site. The incident commander, or another staff member assigned to arrive on the scene, is responsible for making the decisions for scene management and for calling in additional help if required. Any incident site is physically separated into two parts – an inner and outer cordon. When the first responders arrive at the incident site they mount an inner cordon around the rescue zone into which only specially equipped and trained professionals are allowed (Kristensen et al., 2006). One on-site command post is established to control the ER operations and coordinate the interoperation between all of the organisations present including the FRS, police and medical services. The FRS coordinates its own operations within the inner cordon. The medical services coordinate their activities together with the needs of the FRS and the services of the hospitals. The police coordinate their own activities and secure the boundary from access by the public.

ER operations can be classified into three distinct rhythms. The initial rhythm is the mobilisation rhythm in which the fire engines are dispatched by the command centre to the incident site. In this rhythm, the priority of the fire crew is to confirm the information received from the command centre and prepare themselves mentally and physically for the coming actions. The second rhythm is the preliminary situation assessment rhythm starting with the arrival of the fire crew at the incident site and ending with the completion of the preliminary situation assessment. The priority of the fire crew in this rhythm is to decide the tactical mode and to request additional resources. The third rhythm is the intervention rhythm starting when the physical intervention begins and ending at the completion of ER operations. The priority of the fire crew in this rhythm is reducing the loss caused by the disaster and ensuring the safety of the fire crew. There will be some overlap between the preliminary situation assessment rhythm and the intervention rhythm as some initial physical intervention may happen before the preliminary situation assessment is completed or even immediately on arrival. In most cases, police and ambulance services mainly take part in the intervention rhythm. This classification is similar in spirit to the one proposed by Landgren (2006), but with a different definition of the second and third rhythms and the recognition of possible time overlap between these two rhythms.

7.3   GENERAL INFORMATION REQUIREMENTS FOR EMERGENCY RESPONSE

There is rich literature identifying information requirements to meet the needs of first responders in their ER operations. Focusing on a Dutch case, Diehl et al. (2006) investigated user requirements for the work of the different actors in ER including police, fire brigade, ambulance and municipalities and other institutions. They highlighted the importance of getting real time and dynamic information about the crisis and exchanging information between different partners at different administrative levels. Starting with specific user requirements collected in Calabria, de Leoni et al. (2007) presented more general user requirements for supporting communication between control rooms (back-end centres) and on-site rescue teams (front-end teams). Forty-two requirements previously identified in emergency and incident management have been ranked in terms of their priorities by Robillard and Sambrook (2008). A comprehensive list of information requirements of four core members of FRS were given in Yang et al. (2009a,b).

Jackson (2006) summarised information requirements for protecting emergency responders in his speech in the Government Reform Committee, United States House of Representatives. From our research, the following information is considered to be general for the needs on site not only for protecting emergency responders, but also for ensuring the success of emergency operations:

- Information about the environment – When first response teams arrive on site, they have very limited information about the environment, such as the building infrastructure, number of occupants or the exact location of the hazard. Further, they do not know whether the building/underground station is safe to enter or how to most efficiently deal with the hazard (Yang and Frederick, 2006). Many responders may be facing unfamiliar or rare hazards. Incident commanders need to know these hazards and have a plan to deal with these hazards before the commanders deliver their responders to face the hazards.

- Information on response participants – When a disaster occurs, many organisations will be involved in the response. Knowing who is involved in the response, what they are doing, and what resources they are bringing to the scene may give incident commanders information on the responders at the scene and their activities. Sharing these data within individual responder organisations enables cooperation of activities at a response.

- Information on up-to-date casualties – Getting and rapidly sharing the updated casualty information and reporting, including accident location, causes and severity among involved organisations, is critical to ensure responders take appropriately protective measures and to quickly receive emergency medical services during the response. Currently, this information is generally collected within individual organisations and published after the response operation is finished.

- Information about available equipment and other resources – Once a disaster occurs, large amounts of equipment and other resources are quickly delivered to the area by many governmental and non-governmental organisations. Often there is no central control or storage for the equipment. Sharing information about available equipment is critical to ensure responders find what they need from the stocks of equipment at an incident scene. Ensuring that responders have the equipment they need becomes more difficult in the charged and high-pressure atmosphere of an on-going disaster response. Since many different organisations may be involved in managing logistics, a central information management system for tracking what kinds of equipment is in use, where replacement supplies are available and how to match them to meet individual responders' needs are extremely important at a large disaster response.

To further describe and classify user requirements, this section provides a tabular model of the information requirements for ER operations, as shown in Table 7.1. The tabular model presents the information requirements in two dimensions with multiple views. Horizontally, it separates the requirements into two parts: one for front-end teams and another for back-end teams. Front-end teams are directly involved in frontline duties, while back-end teams are located in a command centre that is geographically away from the incident scene. Then, the requirements for front-end teams are further classified into mobilisation rhythm, preliminary situation assessment rhythm and intervention rhythm in terms of the operation stages in which the information are requested. The components in the vertical dimension include the priority of the tasks, category of information, requester of information, source of information, richness of information and importance in real time. For major disasters, the situation might be more complicated than what is described above. For example, some ER organisations may join ER operations directly in the intervention rhythm. Nevertheless, the requirements for front-end teams will be still classified as the ones shown in Table 7.1.

TABLE 7.1
Tabular Model of Information Requirements for ER Operations

	Mobilisation	Front-End Preliminary Situation Assessment	Intervention	Back-End (Command Centre)
Priority of the tasks	Prepare first responders and make sense of what will face them upon arrival	Decide the tactical mode and the request of additional resources	Reduce the loss caused by the accident and ensure the safety of the fire crew	Optimise resource allocation and despatch
Category of information	Environmental conditions	Environmental conditions, available resources	Environmental conditions, information on response participants, status of casualties, available resources	Environmental conditions, information on response participants, status of casualties, available resources
Requester of the information	Incident commanders, fire fighters	Incident commanders	Incident commanders, fire fighters	The command centre
Source of information	Command centre, central database, physical sensors installed in the incident scene, Internet	Local people, physical sensors installed in the incident scene	Local people, on-site officers, physical sensors installed in the incident scene	No emergency personnel, on-site officers, central database, Internet
Richness of information	Low	Medium	Medium	High
Importance in real time	Important	Very important	Extremely important	Less important

Source: Adapted from Yang, L., Yang, S.H., Plotnick, L. 2013. *Technological Forecasting and Social Change*, 80(9):1854–1867.

In the above tabular model, environmental conditions refer to any information about the incident scene such as the building structure, the number of occupants or the exact location of any hazard, the trapped victims and the fire fighters inside the building. Information on response participants include who they are and what expertise they are providing. Status of causalities includes the number of causalities, locations, causes and severity, etc. Available resources may include important equipment, food, medicine and other resources present at the incident scene.

7.4   INFORMATION REQUIREMENTS OF CORE MEMBERS IN EMERGENCY RESPONSE

The four general categories of information about environment, on response participants, on up-to-date casualty information and about available equipment and other resources clearly apply to fire incidents, as a particular type of disaster. This section now focuses specifically on fire incidents in and around large-scale structures and considers the end-user requirements analysis for core personnel who act as first responders in such an emergency situation. These core members in ER are: incident commanders, sector commanders, entry control officers and front-line fire fighters. Other job roles, apart from these core members of the first responder team, are commonly considered to be introduced mainly to maintain the appropriate span of control so that they will reduce both the mental and physical workload of the core members. Therefore, any on-site support system should essentially look after the needs of these four types. The Fire Service Manual (2008) reveals that the requirements of other ER members are a subset generated from the combined requirements of these core members. Therefore, it is reasonable for us to assume that if any on-site information system is able to meet the requirements of these four types of core members, it could easily be adapted to assist any other supporting roles in the fire and rescue operation. The information requirements of different job roles in the fire fighter hierarchy are summarised below.

7.4.1   INFORMATION REQUIREMENTS OF A FRONT-LINE FIRE FIGHTER

The information requirements of the front-line fire fighter include

- Information on the immediate surroundings of the fire fighter, for example, environmental temperature, smoke and CO concentration behind a door, and any possibility of building structure collapse or other critical dangers.
- Information on the fire fighter's body health, including body temperature, rate of breathing, the probability of them getting lost inside a building, running out of oxygen or suffering from extreme exhaustion.
- Information on casualties, including where and how many causalities are trapped inside an accident building, condition of the identified casualties (dead or alive), suitable location for keeping casualties till evacuation.
- Information on other crew members such as body health of fellow fire fighters and their location.
- Overall contextual information on the sector in which they are operating, such as any announcement from a sector commander and information on operational activities being carried out in the vicinity.
- Information on fixed resources and installations around the area of operation, such as sprinklers, ventilation outlets, water drains, fire fighting shafts and wet and dry riser outlets.
- Information on welfare, such as where to find food and choice of food, water, the expected duration of work and arrival of any relief.

- Information on the assigned tasks and resources, including the physical boundary of the assigned tasks, expected result, appropriate personnel protective equipment (PPE) and special equipment available for use.
- Information on any hazard material together with their characteristics and identification.

Their requirements also include the following real-time alarms:

- Out of range alarms
- Health alarms
- Evacuation alarms
- Out of route alarms relevant to an individual front-line fire fighter
- Environmental alarms such as possible back draughts or flashovers

The interviews also highlighted several information needs of the front-line fire fighters, which are considered as essential to provide them with higher levels of situation awareness (SA). These higher level SA requirements can be described as

- Search and rescue route options. Possible routes to search and rescue any trapped people inside a building, routes to the source of the hazard (fire), routes to the items to be salvaged.
- Real-time navigation support. Finding the way out of a potentially danger-ous place is a crucial task for fire fighters especially when working with breathing apparatus that can provide support for only a very limited amount of time. Real-time navigation system supports fire fighters enabling them to fulfil their mission and return safely.
- Alternative route options due to contextual change. If the pre-defined route has been made unusable due to the environment change what is the alterna-tive route to complete the assigned task?
- Dynamic contextual changes along the route: spread of fire and other hazards, occurrence of new risks and hazards.

7.4.2 Information Requirements of Entry Control Officers

Entry control officers are the team leaders of front-line fire fighters. All the infor-mation listed as required by the front-line fire fighters should also be available to an appropriate entry control officer. In addition, entry control officers need some additional management information such as

- Evacuation status of individual front-line fire fighters
- Assigned profiles of individual front-line fire fighters
- Tasks assigned for an entry control officer
- Assigned resources and new resource requests, such as availability of emer-gency back up teams, number of fire fighters to be withdrawn and available backups
- Completed search and rescue efforts

- Status of the ongoing search and rescue efforts
- Suitable entry control points and their locations
- Contamination levels of evacuated fire fighters
- Information on nearby entry control points and the current operations

To protect their front-line fire fighters entry control officers need the following real-time alarms:

- Evacuation failure alarms
- Alarms due to unexpected route changes of the front-line fire fighters
- Duty assignment alarms

7.4.3 Information Requirements of Sector Commanders

A sector commander is in charge of a sector in an incident scene and co-ordinates the response activities of a number of entry control officers who belong to the sector. A sector commander requires the same information, including the real-time alarms as that of an entry control officer. However, more information is required by a sector commander to cover the whole sector and provide a summary of the status of all the front-line fire fighters. In addition, the context forecast and determination of ventilation locations are essential to higher level decision support for the sector commanders. In detail additional information for sector commanders includes

- Assigned physical and human resources to the sector
- Level of work difficulty for officers within the sector
- Resource consumption of essential resource within the sector, such as water and foam
- Information to identify hazard materials and contaminants within the sector, such as relevant hazard material data
- Tactical mode for the sector
- Summary of contextual hazards within the sector, including source, location, level of spread, spread forecast
- Summary of fixed installations within the sector, such as dry and wet riser outlets, sprinklers, fire alarm panels, controls of electricity and gas
- Location of suitable drains for waste water
- Information on availability of expert support for the sector, such as contact and location details of safety officers, paramedics, decontamination officers, location of triage and first aid
- Competence of allocated officers and fire fighters
- Requirement availability and location of welfare, for example, food and water
- Status of physical and human resource requests, for example, location and movement of fire engines, expected time of arrival
- Information on items to be salvaged, such as description of items, ranking of items to be salvaged, handling advice and their location

- Information on decontamination requirements within the sector, for example, level of decontamination and required level of PPE
- Information on construction and structure of the building, for example age and building materials

7.4.4 INFORMATION REQUIREMENTS OF AN INCIDENT COMMANDER

An incident commander is the most senior staff member at the incident scene and is responsible for all the sectors of an incident. The information required by an incident commander includes an incident summary report and access to detail when necessary. The incident commander also needs to be aware of all relevant information about the activities and capabilities of other participating organisations. The incident summary report periodically provides the overall status of the incident, such as the number of casualties being rescued, injuries, deaths, tactical mode changes and sector details. Information for making the initial risk assessment and ranking casualties at the incident was identified as a unique SA need for the incident commander. Additional information for the incident commander on top of sector commander requirements includes

- Information on external water resources, such as public and private pools
- Detailed information on vulnerable buildings around the vicinity of the incident, for example, power plants, petrol stations, schools and hospitals and details of their operations
- Information on overall progress of search and rescue, fire and salvage operations, such as rate of rescue and salvage, increase or decrease of spread of fire
- Information on surrounding domestic population, such as population density and location
- Information on nearby environment to be protected, such as water sources and wild life
- Weather and weather forecast around the incident, especially the wind condition
- Information on traffic arrangements around the incident
- Information on incident terrain, for example, ground slope
- Information on the overall incident hierarchy
- Contact information of incident specific specialists, such as architects, engineers and safety officers
- Information on building occupiers, such as type of occupiers, useful historical behaviour, for example, evacuation reluctance
- Information on sectors, external boundaries and cordons, such as hazard zone, operational safety zone and public evacuation zone

The information requirements of individual core members in the ER team can be organised into a hierarchy as shown in Figure 7.1. This hierarchical structure is evolved from the chain of command of the fire which is a para-military organisation with strict ranks and roles. The rank increases upwards. The arrows pointing

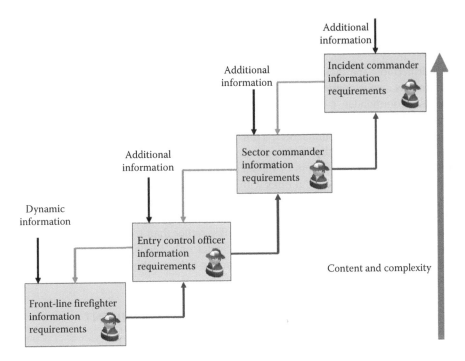

FIGURE 7.1 Information requirement hierarchy for emergency response.

upwards represent the information flow from lower rank members to higher rank members, and the arrows going down show information flow in the opposite direction. Dynamic information is mainly about the immediate surroundings of the fire fighters and is collected from the environment automatically or manually. Additional information refers to any information not received from the higher rank members (e.g. direct commanders) and/or lower rank members (e.g. directly supervised staff). The information requirements of front-line fire fighters form the most basic part in the hierarchy. All the information requirements of front-line fire fighters are fed to entry control officers. Control commands and updated information including control commands from the entry control officers are sent to the front-line fire fighters. Similarly, the entry control officers and the sector commanders feed their information to, and receive the control commands from, their direct commanders. Incident commanders are at the top of the hierarchy and hold all the information about the incident, including the status and capabilities of other participating organisations. This hierarchy organises information requirements in terms of the emergency first responder's goals, rather than presenting them in a way that is technology oriented. This hierarchy also identifies the common constituent of all the end-user requirements, recognises the importance of information sharing among different members and highlights the increasing complexity and content of the information sent from the front-line fire fighter layer to the incident commander layer. This highly hierarchical structure provides the fire crews with the required confidence for making very fast decisions to respond to the emergency. The application scope of this structure is limited to fire brigades only when they are working inside the inner cordon. Other participating organisations who are working outside the inner cordon do not follow this hierarchy.

Furthermore the incident commander is at the top of the hierarchy and directly commands sector commanders. The sector commanders have an up-link with the incident commander, a down-link with their entry control officers and horizontal links with other sector commanders. Similarly, entry control officers have an up-link with their sector commanders, a down-link with their front-line fire fighters and horizontal links with other entry control officers within the same sector. Front-line fire fighters are at the bottom of the hierarchy, having an up-link with their entry control officers and horizontal links with other front-line fire fighters who are under the supervision of the same entry control officers. The content and complexity of information increase from bottom to top. Information sharing happens only through the up-links, down-links and horizontal links, rather than from anyone to anyone. For example, an individual front-line fire fighter only needs to communicate with their entry control officer and other front-line fire fighters who are under the supervision of this entry control officer.

7.5   GOAL DIRECTED TASK ANALYSIS: A NON-RESILIENT APPROACH

The elicitation of end-users' information requirements has always been a crucial part of the information system development cycle. It aims to 'understand what information end users really need from IS to make decision'. Information requirements are the information-related part of end-user requirements. Goal-oriented analysis has been proposed in the requirement engineering (RE) literature to capture the intentionality behind software requirements (Pressman, 2005). Goals are defined as something that stakeholders hope to achieve in the future and are a useful abstraction to represent stakeholders' needs and expectations. A formal goal structure and modelling was defined by Rolland et al. (1998). Goal-based approaches offer a very intuitive way to elicit and analyse requirements (Ali et al., 2010).

Among those goal-based approaches, goal directed task analysis (GDTA) (Endsley et al., 2003a) was developed with the aim of developing systems that enhance the end-users' SA in the aviation and military domains, and has been adopted widely in information requirements gathering, mainly due to their limited need of labour and resources and because they are easy to use.

GDTA focuses on the basic goals of end users, the major decisions that need to be made to accomplish these goals, and the SA requirements for each decision (Endsley et al., 2003b). A decision is the selection between possible actions. Decision making can be regarded as the mental processes resulting in the selection of possible actions to achieve a goal. Every decision-making process produces a final choice. An SA requirement is defined as the end-users' dynamic information needs that are relevant in a particular domain and context rather than the static knowledge the end users must possess. GDTA seeks to determine what information end users really need to achieve each goal. Generally, end users are interviewed, observed and recorded individually. The resulting analyses are pooled and then validated by a large number of end users. The information obtained from these are organised into charts depicting a hierarchy of goals, sub-goals, decisions relevant to each sub-goal and the associated SA requirements for each decision.

7.6 GOAL DIRECTED INFORMATION ANALYSIS: A RESILIENT APPROACH FOR EMERGENCY REQUIREMENT CAPTURING

Goal directed information analysis (GDIA), a resilient approach for emergency requirement capturing was proposed by Yang et al. (2014) based on the above GDTA described previously. GDIA intends to start from physical task identification and then move to a goal structure rather than starting from goal identification such as GDTA. This is because goal modelling is known to present a number of difficulties in practice (Rolland et al., 1998) and is not best suited for use at all for first responders. Scenarios are often used to concretise or discover goals (Rolland et al., 1998). We use scenarios to discover physical tasks and then use those identified tasks to construct a goal structure. We argue that eliciting physical tasks rather than operational goals from first responders is much easier and more meaningful due to the unique features of ER operations. In fact, it is essential to split the whole GDIA approach into a series of clear and repeatable steps with clear and straightforward methods used within each step. We split the task identification into three steps and then employ the GDTA as the later four steps which lead to seven proposed steps in GDIA:

- Step 1: Context discovery
- Step 2: Establish scenarios and select end users
- Step 3: Identify physical tasks
- Step 4: Obtain goals and sub-goals
- Step 5: Goal structure validation
- Step 6: Identify decisions and information
- Step 7: Validate goal-decision information diagrams

These seven steps form a solid framework for the 'elicitation of information requirements of emergency first responders'. The links of these steps are illustrated in Figure 7.2. There are two inter validation-loops for Steps 4 and 5 and Steps 6 and 7. The iteration between Steps 2 and 6 allows further modification happening if necessary.

Step 1: *Context discovery.* Very few of the techniques and tools described in the literature include explicit discussion of the initial steps, but these can be crucial for eventual success, so we propose and explain an explicit first step labelled context discovery. Context discovery is developing an initial understanding of the context being investigated via a few face-to-face discussions with end users and observation of real-time practices and training sessions. Although it was impossible to observe a real large-scale incident, this study sought as many opportunities as possible to conduct field observations in environments which closely resembled an actual fire, such as training practices and simulation exercises. Context discovery seeks to achieve a basic general knowledge of the characteristics, operations, job roles and responsibilities of

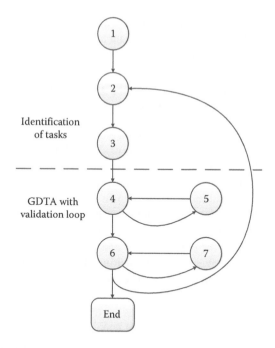

FIGURE 7.2 GDIA process.

the context being explored. This initial understanding is crucial for an IS development practitioner as it helps to build suitable scenarios, select end users for interview and creates confidence to carry out the remaining steps of the approach.

Step 2: *Establish scenarios and select end users.* A scenario is a postulated sequence or development of events which describes a possible story, unfolding from a set of initial conditions to a set of plausible futures. Scenarios are frequently used in emergency planning, preparedness and training. Efforts are always made to ensure the scenarios are as realistic as possible. Suitable scenarios are needed during the interview sessions to overcome the limitations of unstructured interviews. The scenarios are used to probe and challenge the mental models, thought habits and unrecognised assumptions of the end users being interviewed. This enhances the interview focus as well as the imagination of the end users when the investigator is seeking insights into the context. The scenarios could be in the form of paper-and-pencil exercises, tablet digital drawing or computer-supported simulations.

The scenario concept can be presented in a unified way as shown in Figure 7.3. It describes how and what actions trigger the transition from one state of an incident to another. Each state is defined by a set of attributes. For example, an initial state of a fire incident is defined by such attributes as site, location, time, incident type, level of severity. These actions are essentially associated with a set of tasks and enable the state transition. The actions taken in a fire incident could include identifying the fire location, searching people tracked inside, etc. The formation of the intermediate states is decided by the actions that have been taken. The links between individual intermediate states could be in variety like sequence, mesh-link or cyclic.

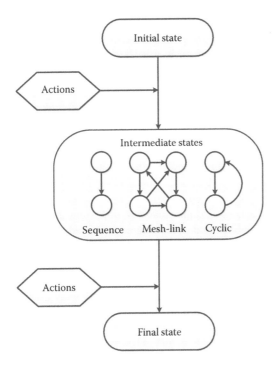

FIGURE 7.3   Scenario construction.

In the sequence link, a state occurs if a previous state has been achieved and their occurrences are in sequence in terms of time. Mesh-link means that a state could be transferred to one of several states depending on the actions having been taken. In the cyclic link, a state may re-occur, for example, a room where fire has been extinguished may be on fire again.

Studies have been reported on exploring how to use scenarios to directly discover requirements (Seyff et al., 2009). We do not take this approach as too many steps are omitted in the transitions from scenarios to requirements.

'Homogeneous sampling' (Patton, 2002) is proposed to identify suitable end-users for conducting interviews and validations. This strategy identifies appropriate end users with a particular job role. Furthermore, it includes identifying any important sub-groups within a group. For example, within any job category it helps to identify the more experienced officers and the novice officers. Homogeneous sampling also involves selecting a particular sub-group of end users where different kinds of participants are included.

Step 3: *Identifying physical tasks.* This step involves a clear identification of the physical tasks (actions) required in the scenarios. This involves the IS development practitioner spending time with end users to discuss each scenario and to clarify any confusion. Thereafter, each selected end user is asked to describe the possible activities and tasks (actions) he (or she) could carry out in each scenario for each state transition.

There are three types of relationships among the tasks during execution, as shown in Figure 7.4. If tasks are carried out one by one they are in a sequence relationship. If tasks are carried out simultaneously, they are in a parallel relationship. If the execution of a task is triggered by an event, such as building collapse, they are in an event-driven relationship.

Task identification should be carried out by one-to-one, face-to-face semi-structured interviews with the selected end users against the scenarios. Semi-structured interviews involve an interview guide supported by various forms of interview probes on topics the interviewer is free to explore (Patton, 2002). This gives the interviewer maximum flexibility while helping to retain focus when time is limited or the interview is interrupted by a real emergency (Hancock et al., 2007). During this step the scenarios are expected to expand the end-users' mental models during task identification so that they do not miss any important tasks nor describe irrelevant tasks. To avoid any bias and to capture complete and accurate requirements, a balance of both experienced and novice end users should be maintained. All this data should be analysed to develop the task structures. The intention of this analysis is not to achieve greater precision but to identify groups of tasks so they can be used together with the scenarios as the specific interview probes in the next step.

Step 4: *Obtaining goals and sub-goals*. Semi-structured interviews are conducted with both novice and experienced end users (preferably with the same end users who described the tasks) to identify goals and sub-goals. With the previously captured task structures, the IS development practitioner can ask specific questions: '*Why do you want to carry out this task, or that task…?*' '*Why do you think this task, or that task… is important?*' '*What goals can be met by conducting these tasks in the scenarios?*'. The scenarios used for the task identification are referred to at the start of the interview and again in the middle to maintain the mental focus of the end users. A combination of both scenarios and tasks provides good guidance to explore the goals with the end users, in contrast to what is likely to happen in the GDTA goal-identifying interviews.

The concept of a goal hierarchy, named as a goal structure here, is taken from GDTA and so it is expected there is a goal structure with primary goals on the top, then secondary goals, and then tertiary goals and possibly even a lower level

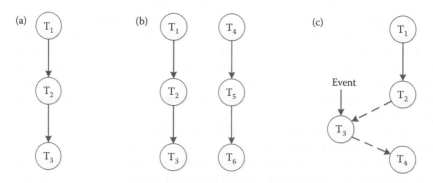

FIGURE 7.4 Three types of task relationships during execution: (a) sequence, (b) in parallel and (c) event-driven.

beneath, in which they all fit together into a goal structure for every particular job role. As shown in the upper part in Figure 7.5, several tertiary goals are achieved to fulfil the secondary goals. A set of primary goals are finally satisfied by means of secondary goals. It must be noticed that it may not be necessary to have tertiary goals or lower level sub-goals for every job role. It should also not be expected that all the goals derived from the scenarios and executable tasks via the interviews are at the lowest level in the goal structure. Actually, the goal fragments elicited from the interviews may be mapped to a goal at any level in the goal structure.

In Figure 7.5, we show the process of generating a goal structure based on the scenarios and tasks. Each end user has a job role boundary. Inside this job role boundary, all the scenarios and executable tasks are used to derive the goal fragments. A goal is identified and composed by examining several relevant goal fragments that have been obtained from end-users' interviews.

We adopted three types of relationships proposed by Rolland et al. (1998) to organise all the identified goals:

- *AND relationship* represents that two or more goals are met together to achieve another complete goal. This AND relationship is used to derive the goal structure in a bottom-up manner.
- *OR relationship* represents alternative ways of fulfilling the same goal. This OR relationship is adopted to identify and remove the redundancies in the goal structure.

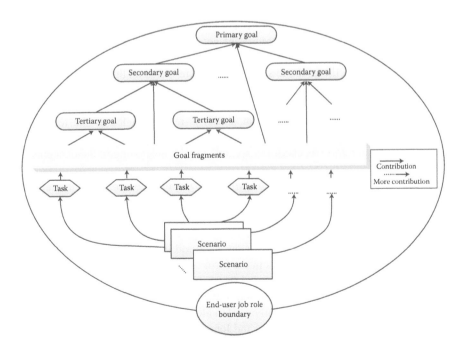

FIGURE 7.5 Process of generating a goal structure.

- *Refinement relationship* represents the abstraction of hiding details of sub-goals in order to focus on essential aspects. Refinement relationship is taken to decompose a goal into a set of sub-goals at a lower level of abstraction. It derives the goal structure in a top-down manner.

Step 5: *Validation of the obtained goals and sub-goals.* It is crucial that the goal structures are sound before moving on to use them in the remaining steps. There is a risk that goals are misinterpreted as tasks or tasks are misjudged as goals in the initial goal structures (Hoffman and Militello, 2008). For example, in fire response, *'have a quick glance at the building plans on the way to the incident site'* is a task of first responders. The associated goal for this task is *'ensure appropriate contextual awareness before reaching the incident'.* Also there is always a risk of missing some of the goals or sub-goals during the initial analysis. Furthermore, there can be terminology mismatches or inconsistencies across the goal structures. Thus, the goal structures developed initially could carry significant errors or be incomplete. It is very important to make sure that these initial goal structures are meaningful, accurate and complete enough to be used as the interview probes in the information capturing phase. The validation process includes the following steps:

- Distribute the initial goal structures among the end users already interviewed.
- Carry out a brainstorming session with end users who were previously interviewed (or meet end users individually) and revisit the goal and sub-goal diagrams with them to minimise initial anomalies.
- Discuss the goal structures with different end users who represent other relevant organisations or regions. This improves the validity of the interview findings so that they represent generic goals of the domain being investigated.
- Finally, update the goal and sub-goal diagrams based on the comments and suggestions.

Brainstorming (Maguire and Bevan, 2002) is recommended as the technique to conduct the validation sessions. Feedback during the validation sessions can be collected in written form as end users are able to add, delete or edit the elements of the diagrams.

Step 6: *Identifying decisions and acquiring information.* At first, with the support of goal probes and the original scenarios, end users are requested to describe the decisions required to achieve each and every individual goal or sub-goal.

This is expected to reveal the decision making required to achieve each goal or sub-goal. Bearing this in mind the decisions captured in this step could be either the actual decisions or other statements in the form of questions, problems or any other issues that indicate the decisions needed to reach the goal. What is more important at this step is to identify as many relevant probes as possible that represent the decisions, so that they can be subsequently used to identify the information requirements of the end users. Endsley et al. (2003b) and Jones and Endsley (2005) have confirmed that the decisions captured with the application of the GDTA could take the form of problems or questions related to the decision. Immediately after the discovery of the decisions required, end users are further probed and requested to describe their information requirements to make such decisions. Figure 7.6 shows the discovering process of decisions and information requirements. Following Endsley et al. (2003a), the information requirements obtained are organised under three headings corresponding to the three increasing levels of SA, where level 1 is mere perception and awareness of facts, but level 2 is comprehension and understanding of the overall current situation and level 3 is projection of the current situation into the future.

Because it is assumed there will normally be a large number of lower level goals, trying to incorporate all the decisions and information requirements into a single hierarchy would be extremely unwieldy. Nevertheless, it is highly desirable to have some form of diagrammatic representation of the decisions and information requirements linked to the goals. Goal–decision–information (GDI) diagrams are proposed as a solution to the problem. Eventually, there needs to be a GDI diagram for each of the lowest level goals. Such a goal might be a secondary goal or a tertiary goal or an even lower level goal. A typical GDI diagram for a tertiary goal (Goal X) is shown schematically in Figure 7.7 as an example. Above Goal X are shown the higher level goals in the hierarchy. Immediately below the chosen Goal X, there is a box indicating the relevant decisions in order to achieve the Tertiary Goal X and there is another box further down in the hierarchy listing information required to enhance the end-users SA to make such decisions. This is designed to ensure that the context is clear and there is no confusion, ambiguity or duplication within the GDI diagrams.

Step 7: *Validation of the obtained GDI diagrams.* GDI diagrams validation is to make sure the GDI diagrams accurately reflect end-users' needs and they are consistent to each other and comprehensive as well. Formally, expressed information requirements can be checked using formal verification techniques such as static analysis or model checking. However, a large number of symbol models are required before these formal validation techniques are applied, which obviously limit their use (Pitula and Radhakrishnan, 2011). This study suggests using informal techniques such as walk-through, reviews and checklists for the validation purpose. In all the cases, the quality of GDI diagrams is assured by having end users reviewing and approving of the diagrams.

At first, the diagrams are sent to the end users who participated in the interviews with instructions on how to interpret them, along with a request to identify any

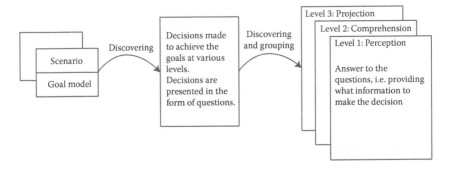

FIGURE 7.6 Discovering process of decisions and information requirements.

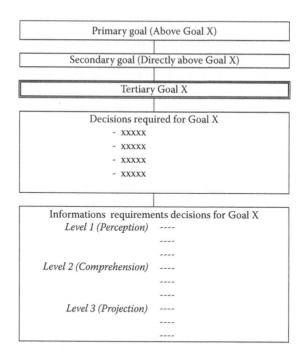

FIGURE 7.7   Schematic representation of GDI diagram for a Tertiary Goal X.

missing information or errors. Thereafter, the diagrams are distributed among other end users who have not previously participated but who are sufficiently familiar with the domain, perhaps from other areas of the county. The feedback obtained from these various sources of validation is integrated and considered. Such integrated feedback is then used to make the necessary modifications to the initial GDI diagrams. The resulting GDI diagrams are expected to be more accurate and complete after incorporating such feedback. The comprehensive information requirements are presented in the final GDI diagrams.

7.7   APPLY GDIA FOR INFORMATION REQUIREMENTS CAPTURING

The proposed GDIA method, as well as the associated methods, was tested in the United Kingdom government funded fire safety project, SafetyNET (Yang et al., 2013), which aimed to prove the concept of an information platform for on-site fire ER operations (Yang et al., 2009b). This project focussed on supporting emergency responders to fire incidents in the building environment. In particular, the concern was with large building structures, such as high rise residences or other multiple occupancy buildings, office blocks, shopping malls, factories and warehouses, but did not include individual houses or fires in the open, such as forest fires. These sorts of incidents are referred to as 'large-structure fire emergencies' by the U.K. FRSs and there are general procedures laid down for dealing with such incidents. The purpose of the proposed information platform in this test was to provide on-site support to the emergency first responders called out to such an incident.

7.7.1 CONTEXTUAL UNDERSTANDING

In this case study, the context discovery explorations started with references to operational procedures, policy and service documentations of the U.K. FRSs. Some early observations of several emergency first response training sessions and initial discussions conducted with fire and rescue officials were also helpful to strengthen the contextual understanding. At this point the documents and initial discussions were also used to discover the similarities and differences between the three local FRSs selected for this study: Leicestershire FRS, Nottinghamshire FRS and Derbyshire FRS. As the SafetyNET system focused on providing on-site decision support for the emergency first responders, the on-site incident command structure was chosen as the focus of the test study.

Usually the incident command structure operates in a temporary control room located at the emergency site. Furthermore, the context discovery revealed that, in these circumstances, there are four core end users, namely: incident commanders (IC), sector commanders (SC), breathing apparatus entry control officers (BAECO) and breathing apparatus wearers (BA wearer). These BA wearers operate the equipment and tackle the incident directly, so they are also referred to as the front-line fire fighters. The identification of the IC, SC, BAECO and BA wearers as the four core job roles is supported by the following two interview excerpts:

One of the station managers in the Derbyshire FRS said *'If you want to completely understand the needs of fire fighters during their operations you better talk to people who work as BAs, BAECOs, operational SCs and the ICs. These jobs are the main links of our chain of command. You simply can't survive without the support of these four jobs during any type of bigger incident. Needs of other jobs are similar to the needs of these four. Actually you can say it is a subset'.*

One of the senior officers of the Command Support Team Developments in Leicestershire FRS said *'If you genuinely want to develop a system supporting us during large incidents my advice for you is to separately understand the needs of the different jobs. You must not miss IC, SC, BA and BAECO jobs. Even if you only managed to cover these four I can accept you have covered almost 99% of our requirements during operations at the incident site. Requirements of other jobs can be part of either one of the above or several of them'.*

7.7.2 ESTABLISHING SCENARIOS

After initial familiarisation of the operational context, incident sites were selected to build suitable scenarios. Four incident sites within the boundary of Derbyshire FRS were selected and four associated scenarios were defined. Each scenario consists of the attributes of the initial state of the incident like the incident site, the location of the fire and start time with a detailed sequence of the intermediate states and the final state of the events built in the form of a story. The summary information of these initial states is shown in Table 7.2.

TABLE 7.2
Initial States of Four Established Scenarios

Incident Site	Location of the Fire	Time of the Incident
Westfield Shopping Centre, Derby	One of the clothing shops on the 2nd floor of the Westfield Centre	Weekday at 3:00 p.m.
Rolls Royce Nuclear Facility, Derby	Rolls Royce Nuclear Building	Weekday at 10:00 p.m.
Bath Street Community Housing, Derby	One of the flats on the 4th floor of Bath Street Flats Community Housing	Weekday at midnight
Derbyshire Royal Infirmary Nurses Quarters, Derby	One of the nurses quarters on the 5th floor of Derbyshire Infirmary	Sunday at 5:00 p.m.

Several field visits to, and live simulation exercises at, these four sites were arranged as a means of getting a better understanding of high-risk fire sites in the county. Being a state-of-the-art modern building, the Westfield Shopping Centre was recognised by most of the fire officers interviewed as the best choice for the principal scenario. However, the Westfield scenario has some limitations in covering some of the fire and rescue objectives, especially related to the environment and community, including risks related to chemicals or radioactive material. Therefore, it was suggested that a secondary scenario such as an incident at the Rolls Royce Nuclear facility should also be used. Furthermore, the highly compartmentalised Royal Infirmary nurses quarters building was expected to introduce the dimension of fires in high-rise apartments occupied by well-drilled and disciplined tenants. Similarly, the Bath Street flats as a scenario was expected to bring the dimension of a community condominium occupied by less fire-drilled tenants including families, residents with disabilities and elderly tenants. Together, these four scenarios were expected to cover the overall operational scope of the Derbyshire FRS in relation to a fire incident in a large high-risk building.

7.7.3 TASK ELICITATION

The task identification phase of GDIA was successfully accomplished with eight interviews. During the interviews the four scenarios previously created were used as the main prompting device to elicit the tasks of the four core job roles. Capturing tasks is a straight forward process, compared to other steps in GDIA, since fire fighters are much more familiar with their operational tasks than any other aspect of their operations. The task gathering interviews were conducted with a small number of end users who were very familiar with the incident scenarios. Four experienced and four novice fire fighters representing two appropriate participants for each job role were identified as suitable interviewees. Novice fire fighters are those qualified for their current role but with little experience in that role. However, they had substantial experience of the chosen scenarios from training sessions and their experiences in previous roles.

Task elicitation was based on the constant comparison technique (Boeije, 2002). Interview transcripts were examined to identify task fragments. A task fragment can be a word; a set of words, a complete statement or set of statements that represent some aspect of a significant task for a particular job role during an incident. In this analysis, the conversation parts of each interview were separated out as task fragments. The second activity in the task elicitation process consisted of interpreting the parts as a whole by connecting the matching task fragments to form a unique task of a particular fire fighter job role. For example, two task fragments obtained from two incident commanders are

> I have to always think about the *resources*. I have to *identify the new resource* need and therefore, need to *order them* through my command support team';
>> One of my other priority tasks would be to *evaluate the available resources* and *identify the new requirements* to deploy the operational plans. Also I am responsible for the task of *ordering of such resources.*

The task elicitation process generates the following three unique tasks for the incident commander job role:

- Resource assessment
- Identify new resource needs
- Request new resource needs via command support

7.7.4 GOAL ELICITATION

7.7.4.1 Goal Fragments Capturing

Goal elicitation was carried out by face-to-face interviews with the four core types of fire fighters. Twenty officers, five of whom represented each of the four core job roles, were interviewed. Participants within each job role were selected in such a way that they represented both experienced and novice categories, as well the full-time and retained categories, and covered both urban and rural areas. For example, the following three goal fragments on the operation priorities were captured during the IC interviews through the scenarios and the associated tasks.
Fragment 1:

> In a highly calculated way, fire fighters
>
>> - Will take some risk to save saveable lives.
>> - May take some risk to save saveable property.
>> - Will not take any risk at all to save lives or properties that are already lost.

Fragment 2:

> If you ask a priority, I would first look for the safety of my crews, then saving lives of other general public, next comes the building and rest of the property and valuables, finally environment. Of course we consider environment as a very important concern.

Fragment 3:

Health and safety of people and my team followed by physical and environmental property comprise our topmost aim. Any other things will have to follow these priorities.

These three fragments were mapped to become a primary goal of the IC *'Assure health, safety and welfare of humans, property and environment'* by interpreting the meaning of these three fragments.

7.7.4.2 Goal Elicitation

The goals were identified for each job role by examining the goal fragments that have been captured. AND, OR and refinement relationships among the obtained goals are employed to link the goals together and then finally construct a goal structure for every particular job role.

Figure 7.8 shows this primary goal and its eight immediate sub-goals (secondary) derived from other goal fragments that are associated with these eight sub-goals. There are two relationships in Figure 7.8 in which refinement relationship is from the primary goal to the individual secondary goals, and AND relationship in the opposite direction (bottom-up). This elicitation process is carried out on and on until no refinement relationship is found. For example, the secondary goal *'Ensure appropriate commitment as an IC'* was decomposed into the tertiary and lower level goals in the same way based on other goal fragments. The completed sub-goal structure of this secondary goal is shown in Figure 7.8.

7.7.5 GOAL VALIDATION

Because of time and accessibility constraints, the validation was conducted by face-to-face brainstorming sessions. The participants in the brainstorming sessions were

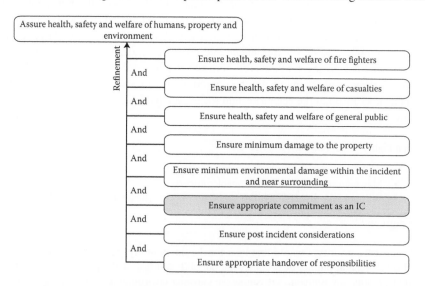

FIGURE 7.8 Primary and secondary goals of an IC.

selected in such a way that they either had real-time experience in large-scale high-risk fires or were expert trainers in that particular job role. Ten validation sessions were carried out for both the IC and SC job roles. Three validation sessions were carried out for each of the BAECO- and BA-wearer job roles. Each participant was met with twice, as all the participants were new to the exercise. In the first meeting, each participant was introduced to the goal structures and any questions and concerns about the presentation format or terminology were clarified. Then they were asked to go away and go through the goal structures and familiarise themselves with the goal descriptions by themselves. Later, a second meeting was held to capture their feedback on the goal structures. During these brainstorming sessions, participants were invited to express their feedback on the goal structures freely. Participants were able to provide comprehensive feedback since they had been given enough time to familiarise themselves with the goal structures prior to each brainstorming session. When necessary, the scenarios were used to probe the participants. All the brainstorming sessions were tape recorded. These unstructured brainstorming sessions allowed the participants to suggest many other goal-related matters as well as to identify what to include and what to delete from the goal structures.

As an example, Figure 7.9 shows the improvement of the secondary goal *'Ensure appropriate commitment as an IC'*. The validation interviewees suggested adding two sub-goals to this secondary goal to detail the contents at a lower level of abstraction. So refinement relationship is employed. These two new sub-goals are shown with dotted lines in Figure 7.9. In other cases, the validation led to the removal of some sub-goals and/or the rephrasing of others.

7.7.6 Decision and Information Requirements Elicitation

7.7.6.1 Data Capturing

Decisions required and the information requirements to make such decisions were also captured by having one-to-one, face-to-face interviews. In total 20 officers, five from each of the four job roles, were interviewed. Before commencing the interviews, arrangements were made to send the scenario outlines to all the participants. This enabled the interviewees to become familiar with the scenarios prior to the actual interview sessions. During the interviews, with the aid of the prepared goal structure and interview guide, end users were asked to describe the decisions they have to make to successfully achieve their expected goals. They were then further probed to spell out the information they required to improve the way they make these decisions.

7.7.6.2 Elicitation of Decisions and Information Requirements

The transcripts of each interviewee were examined to identify relevant fragments. These were carefully assigned to the relevant goal or sub-goal. The constant comparison method (Boeije, 2002) was used as the analysis tool. For each goal of a job role, the transcripts of the most experienced end users were examined first. Thereafter, the transcripts of the other participants were examined. The fragments from each interview were tabulated in parallel in two separate tables, one for the decisions and one for the information requirements. Subsequently, the transcripts were further

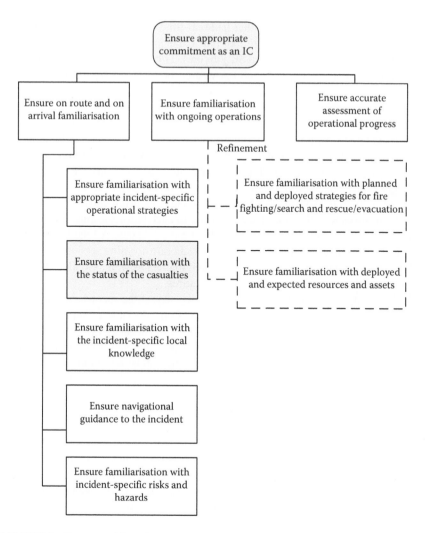

FIGURE 7.9   Decomposition of a secondary goal.

examined for any additional fragments missed in the first round of comparison. This constant comparison technique was repeated for all the goals and sub-goals identified for a particular job role.

Finally, the decisions and the information requirements were combined with the goals to form the 'GDI diagrams' representing a goal, decision and information requirements hierarchy for every job role. The information requirements were further classified according to the three increasing levels of SA: level 1: perception, level 2: comprehension and level 3: projection as described by Endsley (2003b).

The corresponding GDI diagram for that sub-goal is structured as shown in Figure 7.10. For this particular sub-goal the decisions captured were in the form of questions related to the underlying decisions rather than the decisions themselves.

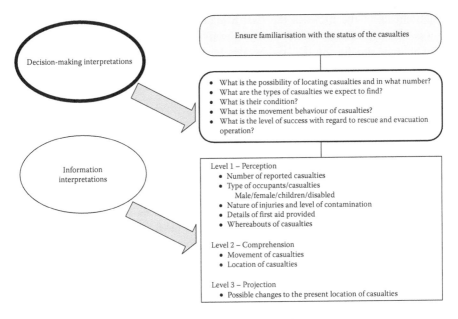

Decision-making interpretations

Ensure familiarisation with the status of the casualties

- What is the possibility of locating casualties and in what number?
- What are the types of casualties we expect to find?
- What is their condition?
- What is the movement behaviour of casualties?
- What is the level of success with regard to rescue and evacuation operation?

Information interpretations

Level 1 – Perception
- Number of reported casualties
- Type of occupants/casualties
  Male/female/children/disabled
- Nature of injuries and level of contamination
- Details of first aid provided
- Whereabouts of casualties

Level 2 – Comprehension
- Movement of casualties
- Location of casualties

Level 3 – Projection
- Possible changes to the present location of casualties

FIGURE 7.10   GDI diagram for a lowest level goal of an IC.

However, these questions were quite effective as probes to obtain the comprehensive information requirements corresponding to the goal. This form of decision-making interpretations was supported by the statement in GDTA – *'Decisions are posed in the form of questions, and the subsequent SA requirements provide the information requirements to answer those questions'* (Endsley et al., 2003b, p. 75). For example, one of the decision questions was *'What is their possibility of locating casualties and in what number'?* The answers to this question were *'number of reported casualties'*, *'whereabouts of casualties'*, and *'location of casualties'*. These were the captured information requirements and were arranged into three increasing levels of SA (Yang et al., 2009a).

7.7.7   VALIDATION OF GDI DIAGRAM

Validation started with the distribution of the GDI diagrams to all the participants of the GDIA interview sessions. Providing feedback on all the GDI diagrams is an impossible task for an individual participant, due to the significant amount of information to be verified for each set of GDI diagrams. Therefore, it was decided to select a random set of GDI diagrams relevant to one particular job role so that the number of selected diagrams could be comfortably discussed in a reasonable time. The brainstorming sessions were conducted. The comments from these sessions were summarised to form a list of comments. Separate sets of proposed improvements were identified for the four job roles. These suggestions were incorporated into the respective GDI diagrams to obtain the validated diagrams. With such rigorous feedback, these validated GDI diagrams were taken to represent the information requirements of the FRSs in the East Midlands Region in the United Kingdom. Altogether, 310 validated GDI diagrams were obtained to cover the four job roles, with 88 diagrams for the IC, 66 diagrams for the SC, 62 diagrams for the BAECO and 107 for the BA wearer.

7.8   DISCUSSION

The main contribution of this study is the set of proposed goal-directed information analysis (GDIA) steps with the methods for their implementation, including the models of scenario, task and goal and discovering process of information requirements. GDIA is a scenario-based, semi-structured interviewing technique, which consists of seven clearly defined and specified steps. GDIA overcomes several application and resource constraints and ambiguities in existing cognitive task analysis tools (CTA). It reduces the risk of non-elicitation or misinterpretation of information requirements possible in most of the other CTA tools. With the proposed validation steps, GDIA should yield comprehensive and accurate information requirements in a fewer number of iterations and interviews compared to other similar approaches. It could be an ideal approach for a single practitioner who has limited resources but is requested to explore SA requirements in emergency related domains.

The case study testing results show that GDIA did fulfil these expectations and made an important contribution to the successful elicitation of information requirements. The test activities involved over 1000 contact hours in the form of face-to-face, semi-structured interviews and brainstorming sessions with fire fighters, together with observations of related training and simulation sessions. A comprehensive set of information requirements for the four core job roles of the U.K. FRSs during a fire emergency operation was represented in 310 GDI diagrams which were fully validated by the end of the testing. It is particularly worth noting that this approach succeeded even though several interviews were interrupted due to officers being called away to attend emergency incidents. These interviews were all successfully completed at a later stage. With some less structured approaches this would have been very difficult.

Rather than just obtaining a list of information requirements, GDIA has led to clear links between the information requirements, the decisions needed and the goals of the four core job roles during emergency operations. It also revealed the similarities and differences between the information requirements of the job roles. The feedback obtained from fire fighters indicated the process had been very successful. This is the first known occasion when the information requirements of different fire fighter job roles, including the requirements of both novice and experienced fire fighters, have been comprehensively identified.

The presented GDIA method is generic, but the test case study focussed on the U.K. fire ERs. Three limitations have been recognised in the test case study. The first limitation is that a relatively small number of scenarios were used. It is possible that adding more scenarios would improve the comprehensiveness of the information requirements. Future research needs to investigate the optimum number of scenarios to use. Secondly, the investigations of this test study were based on a convenience sample of easily accessible local respondents. Developing the necessary rapport with emergency respondents further afield could be more difficult and the establishing of the optimum number of respondents at each stage in other situations may be difficult. Thirdly, of course, the proposed GDIA method was tested on a case study focused on U.K. fire ERs. However, there is no reason to think that proposed generic GDIA method could not be applied to other situations elsewhere, provided there is willing cooperation of the relevant emergency responders.

7.9 CONCLUSION

On-site emergency information systems are needed to provide information on environments, casualties, response participants and available resources that will allow incident commanders to make accurate decisions for efficient ER. It might be easy to collect some information requirements from fire fighters by simply talking to them or referring to their relevant operation documents, but it is never easy to capture accurate and comprehensive information requirements. This chapter has identified the four key end users of on-site emergency information systems as incident commanders, sector commanders, entry control officers, and front-line fire fighters. Each of these end users has unique information requirements that have been explored. A systematic resilient approach which can be adopted by a IS development practitioner has to be employed to elicit accurate and comprehensive information requirements. Moving on from GDTA that is a cognitive task analysis tool, this chapter presents a resilient approach named GDIA. This starts with physical task elicitation rather than goal elicitation. It has seven clearly defined and repeatable steps.

The case study tested the primary data gathering and analysis of GDIA by attempting to elicit a comprehensive set of information requirements for fire fighters responding to a fire emergency operation. Using the proposed GDIA method, 310 GDI diagrams were elicited and later validated. These GDI diagrams present the information requirements of fire fighters in the four core fire fighter job roles. This is the first known occasion that the comprehensive information requirements of fire fighters have been identified, codified and validated. The validation exercises gained a very positive response from a range of frontline fire fighters indicating the process had accurately and comprehensively identified their information requirements, and this was only possible through the application of the proposed GDIA method.

The presented GDIA approach was deployed in a U.K. national fire ER research (SafetyNet, http://www.lboro.ac.uk/microsites/enterprise/safetynet/) and has been used in the design of an emergency decision support system for mass evacuation in New Zealand (Javed et al., 2010). Even though the case study presented in this chapter is fire incident focused, the seven steps in the GDIA are likely suitable for other kinds of emergencies beyond fire incidents because of the generality of the approach.

REFERENCES

Ali, R., Dalpiaz, F., Giorgini, P. 2010. A goal-based framework for contextual requirements modelling and analysis. *Requirements Engineering*, 15:439–458.
Boeije, H. 2002. A purposeful approach to the constant comparative method in the analysis of qualitative interviews. *Quality and Quantity*, 36(1):391–409.

Berrouard, D., Cziner, K., Boukalov, A. 2006. Emergency scenario user perspective in public safety communications systems. *Proceedings of the 3rd Information Systems for Crisis Response and Management Conference (ISCRAM2006)*, Newark, NJ, USA, pp. 386–396.

Danielsson, M. 1998. The cognitive structure of decision making tasks in major versus minor emergency responses. In: Scott, P.A., Bridges, R.S. and Chasterrs, J. (eds), *Proceedings of the Global Ergonomics*. Elsevier: Cape Town, South Africa, pp. 627–632.

de Leoni, M., de Rosa, F., Marrella, A., Mecella, M., Poggi, A. 2007. Emergency management: From user requirements to a flexible P2P architeture. In Van de Walle, B., Burghardt, P. and Nieuwenhuis, C. (eds.) *Proceedings of the 4th Information Systems for Crisis Response and Management Conference ISCRAM2007*, Delft, the Netherlands, Brussels University Press, pp. 271–279.

Diehl, S., Neuvel, J.M.M., Zlatanova S., Scholten, H.J. 2006. Investigation of user requirements in the emergency response sector: The Dutch case. *Second Symposium on Gi4DM*, Goa, India, CD ROM, 6pp.

Endsley, M.R., Bolstad, C.A., Jones, D.G., Riley, J.M. 2003a. Situation awareness oriented design: From user's cognitive requirements to creating effective supporting technologies. In: *Proceedings of the Human Factors and Ergonomics Society 47th Annual Meeting*, Denver, CO, 13–17 October, pp. 268–272.

Endsley, M.R., Bolte, B., Jones, D.G. 2003b. *Designing for Situation Awareness: An Approach to Human-Centered Design*. Taylor & Francis: London.

Fire Service Manual. 2008. 3rd edition, *Vol. 2 – Fire Service Operations*, Crown Copyright 2008: UK.

Hancock, B., Windridge, K., Ockleford, E. 2007. *Introduction to Qualitative Research*. Trent RDSU Focus Group: Leicester, UK.

Hoffman, R.R., Militello, L.G. 2008. *Perspectives on Cognitive Task Analysis: Historical Origins and Modern Communities of Practice*. Psychology Press, Taylor & Francis Group: NY.

Jackson, B.A. 2006. Information sharing and emergency responder safety management, http://www.rand.org/pubs/testimonies/2006/RAND_CT258.pdf.

Javed, Y., Norris, T., Johnson, D. 2010. A design approach to an emergency decision support system for mass evacuation. *Proceedings of the 7th International ISCRAM Conference*, Seattle, USA.

Jones, D.G., Endsley, M.R. 2005. Goal-directed task analysis. In: Hoffman, R.R., Crandall, B., Klein, G., Jones D.G. and Endsley, M.R. (eds.), *Protocols for Cognitive Task Analysis*. Advanced Decision Architectures Collaborative Technology Alliance, US Army Research Laboratory. Technical rept. Florida Institute for Human and Machine Cognition Inc.: Pensacola, FL.

Klann, M. 2008. Tactical navigation support for firefighters: The lifeNet Ad-Hoc sensor-network and wearable system. *Proceedings of the Second International Workshop on Mobile Information Technology for Emergency Response (Mobile Response 2008)*, Bonn, Germany, pp. 41–56.

Kristensen, M., Kyng, M., Palen, L. 2006. Participatory design in emergency medical service: Designing for future practice. *Proceedings of ACM Conference on Human Factors in Computing Systems*, Montreal, Quebec, Canada, pp. 161–170.

Landgren, J. 2006. Making action visible in time-critial work. *Proceedings of ACM Conference on Human Factors in Computing Systems*, Montreal, Quebec, Canada, pp. 201–210.

Maguire, M., Bevan, N. 2002. User requirements analysis: A review of supporting methods. In: *Proceedings of the IFIP 17th World Computer Congress*. Montreal, Canada, 25–29 August, pp. 133–148.

Mehrotra, S., Butts, C., Kalashnikov, D., Venkatasubramanian, N., Rao, R., Chockalingam, G., Eguchi, R., Adams, B., Huyck, C. 2004. Project rescue: Challenges in responding to the unexpected. *SPIE*, 5304:179–192.

Patton, M.Q. 2002. *Qualitative Research and Evaluation Methods*, 3rd edition. Sage Publications Ltd.: London.

Pitula, K., Radhakrishnan, T. 2011. On eliciting requirements from end-users in the ICT4D domain. *Requirements Engineering*, 16:323–351.

Pressman, R.S. 2005. *Software Engineering: A Practitioner's Approach.* McGraw-Hill: Boston.

Robillard, J., Sambrook, R.C. 2008. USAF Emergency and Incident Management Systems: A Systematic Analysis of Functional Requirements. *White paper for United States Air Force Space Command*, 2008, Electronic Text at, http://www.uccs.edu/~rsambroo/Research/EIM_REQS.pdf (last accessed in February 2012).

Rolland, C., Souveyet, C., Achour, C.B. 1998. Guiding goal modelling using scenarios. *IEEE Transactions on Software Engineering*, 24(12):1055–1071.

Seyff, N., Maiden, N., Karlsen, K., Lockerbie, J., Grunbacher, P., Graf, F., Ncue, C. 2009. Exploring how to use scenarios to discover requirements. *Requirements Engineering*, 14:91–111.

Wang, J., Rosca, D., Tepfenhart, W., Milewski, A., Stoute, M. 2008. Dynamic workflow modeling and analysis in incident command systems. *IEEE Transactions on Systems, Man and Cybernetics, Part A*, 38(5):1041–1055.

Wang, J., Tepfenhart,W., Rosca, D. 2009. Emergency response workflow resource requirements modeling and analysis. *IEEE Transactions on Systems, Man and Cybernetics, Part C*, 39(3):1–14.

Yang, L. 2007. On-site information sharing for emergency response management. *Journal of Emergency Management*, 5(5):55–64.

Yang, S.H. and Frederick, P. 2006. SafetyNET/ a wireless sensor network for fire protection and emergency responses. *Measurement + Control*, 39(7):218–219.

Yang L., Prasanna, R., King, M. 2009a. Situation awareness oriented user interface design for fire ER. *Journal of Emergency Management*, 7(2):65–74.

Yang, L., Prasanna, R., King, M. 2009b. On-site information systems design for emergency first responders. *Journal of Information Technology Theory and Application*, 10(1):5–27.

Yang, L., Prasanna, R., King, M. 2014. GDIA: Eliciting information requirements in emergency first response. *Requirements Engineering*, February: 1–18.

Yang, L., Yang, S.H., Plotnick, L. 2013. How the internet of things technology enhances emergency response operations. *Technological Forecasting and Social Change*, 80(9):1854–1867.

Section III

Practice

This section aims to be more practitioner focused as it presents thinking on resilience in a broader context. In Chapter 8, Siemieniuch et al. discuss the topic from a pure engineering systems perspective by exploring systems and also systems of systems. The authors include a number of examples from the real world from which they extract lessons that can guide the building of resilience into systems. Interestingly, the authors see organisation and cultural elements such as leadership, IT systems and organisational structure as key to building resilience. In Chapter 9, Palin discusses the hot topic of supply chain resilience. Palin brings a considerable historical perspective to modern day issues of global logistics, risk and resilience. Here again are multiple examples that prompt thinking about the nature of resilience in today's deeply integrated and global economy. In Chapter 10, Kachali et al. consider the critical phase in resilient systems of post-disaster recovery by looking at the case of the New Zealand earthquake of 2010. The authors conducted a survey that captured the initial impacts and perceptions of organisations that were affected by the disaster. It is interesting to note that Kachali et al. found both challenges and opportunities from such an event – depending on the nature of your business. For example, ICT firms found it difficult to meet the increased demand for their services as organisations could not effectively function without this capability.

8 Supply Chain Resilience
Diversity + Self-Organisation = Adaptation

Philip J. Palin

CONTENTS

Supply has been unchained: Improvements in transportation – highways, fast, high-capacity planes, intermodal trains, trucks and ships – combined with a revolution in computing and communications have transformed a lattice of overlapping supply chains into a shared network for delivery-on-demand.

Only 30 years ago a major retailer often owned most of the links in a chain that connected producer to consumer. No more. Almost everyone specialises in a horizontal or oblique or nodal niche. Each niche forms around functional expertise and comparative advantage. This global commons makes possible more goods at lower cost and with better assurance of quality than ever before. In the last generation, we have experienced a rate of change and improvement in moving goods not equalled since steam-power transformed maritime shipping and made railroads possible.

On 26 June 1974 at a Marsh supermarket in the middle-American state of Ohio, a pack of Wrigley's Juicy Fruit chewing gum became the first retail product sold using a scanner and universal product code (UPC). The use of the UPC and other 'bar codes' allows the supply chain to be digitally monitored, mapped and managed as never before. Logistics has become one aspect of a constantly shifting supply and demand stream.

Increasingly these processes – and the rich information and management resources they make possible – ensure effective, timely and comparatively friction-free transactions between companies and nations. The ability to share digital information in very close to real time has transformed the modern supply chain from supply–push to demand–pull.

Farmers, miners and fishermen still matter. Processors, truckers, wholesalers and retailers still play crucial parts. Ports, railways and highways are still required.

Physical stuff of all sorts still has to move from point A to B (and usually on to points C, D and Z). But at least in the United States, Europe and the Pacific Rim, the digital signals that are sent along largely determine when and where product arrives.

When the strategic capacity for generating demand–pull information persists, the supply chain is very resilient. But disruption or corruption of this information stream also presents unprecedented challenges to supply, especially in crisis situations. In the aftermath of the 11 March 2011 earthquake and tsunami in Japan demand signals went dark across the hardest hit areas of northeast Japan. At the same time demand spiked in Tokyo and other areas far from the impact zone. The supply chain responded adroitly to hoarding behaviour by those whose demand could be communicated with a quick electronic scan. But this same behaviour reduced the capability of Japanese producers and distributors to respond to the critical needs of those who had been rendered digitally mute (Gilligan 2011).

8.1   UNPRECEDENTED HISTORICAL TRANSITION: COMPLICATED TO COMPLEX

Until little more than a century ago, most humans harvested the food that fed them and supplemented their diets, clothing and household goods by trading with near-by neighbours. Long-range trade existed – the Silk Road, Sahara caravans, Indian Ocean pepper traders and more – but this was a high-risk, high-profit luxury exchange. Essential commodities were locally sourced because there was too much risk of long-term disruption.

At the height of their powers Rome, Athens and a few others had sufficient wealth and maritime dominance to depend on distant sources of grain. Tacitus complained, 'The life of the Roman nation has been staked upon cargo-boats and the vagaries of seafaring' (Tacitus Publius Cornelius, at 100 AD). And as Tacitus fretted, this dependence, undertaken in strength, amplified weakness in decline. From the fall of imperial Rome to the British Importation Act of 1846 both policy and practice required that diverse staples be produced as close to home as possible.

This is no longer true. Since at least 1846, production of every sort has trended towards specialisation and the search for comparative advantage. As steam power reduced the cost and increased the volume of trade – and as western navies both projected and protected free trade – the prophecies of Adam Smith and David Ricardo were made manifest in British textiles, German chemicals, Argentine beef and American cotton.

Specialisation begets concentration which begets even greater specialisation which deepens comparative advantage and encourages further specialisation and so the cycle runs. At least so it seems, based on a little less than 200 years' experience.

Steam replacing sails and railways spanning continents made transportation of goods possible where it had previously been nearly impossible. But the cost of transportation and distribution was a very significant aspect of the total cost of goods. Writing of the U.S. economy in 1941, Chester W. Wright, explained,

Of the $65.6 billion representing the total cost of producing and distributing goods, it is roughly estimated that **59 percent or $38.5 billion represents the total cost**

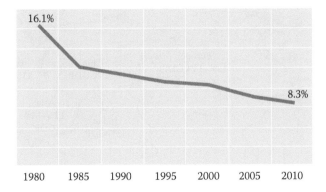

FIGURE 8.1 Logistics cost as a percentage of gross domestic product. (From Bureau of Economic Analysis, 2012, *National Income and Products Accounts*, Washington, DC.)

of distribution... This means that in general it costs the country about 50 percent more to get goods distributed than it does to get these same goods produced. (Wright 1941, p. 775)

In 2011, the total cost of distribution was 8.5 per cent of gross domestic product (see Figure 8.1). What happened? A great deal, but the biggest shifts in these 70 years are related to trucks and telecommunications and computing.

In 1930 the railroads carried approximately 74 percent of the total of domestic intercity freight traffic, while motor carriers carried less than 4 percent... In 1946, the railroads hauled only 43.5 percent of the inter-city freight traffic... In the same year motor carriers had increased their share to 22.7 percent. (Wright 1941, p. 67)

Today rail moves roughly 25 per cent of U.S. freight. Trucks now carry 70 per cent. The already strong trend towards trucking was accelerated by the Federal Aid Highway Act of 1944 and, especially, the Federal Aid Highway Act of 1956 that funded sustained construction of the U.S. interstate highway system. State and federal highway construction has significantly reduced costs and increased competition in the freight and distribution industries, essentially giving the supply chain a significant public sector subsidy. The deregulation of the U.S. trucking industry by the 1980 federal Motor Carrier Act further facilitated innovation and market responsiveness.

Long-haul trucking is highly competitive and an effective cost competitor with rail, water and air freight. Compared with other common carriers, trucking has lower barriers to entry and can flexibly respond to a variety of market shifts. But an expanded number of roads, trucks and drivers cannot claim full credit for the steep reductions in distribution costs that took place in the second half of the twentieth century.

As late as 1980 the cost of logistics as a proportion of U.S. GDP was still more than double what it is today (Bureau of Economic Analysis 2012). Over the last three decades, U.S. supply chains have become something very different than ever before, generating a considerable comparative advantage for the U.S. economy (see Figure 8.2).

For most of human history anticipating demand has been a guess. In a few cases, historical information might be used to project demand and some were

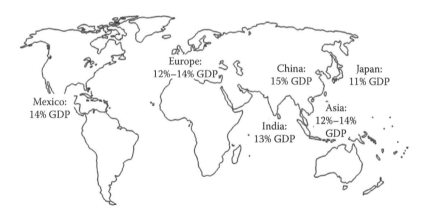

FIGURE 8.2 Comparative cost of logistics as a percentage of gross domestic product. (From Council of Supply Chain Management Professionals, 2011. With permission.)

better – or luckier – guessers than others. But mostly demand was supplied (or not) by producing and distributing what could be supplied at a price the producer hoped the market would pay. A nineteenth century economic doctrine, as stated by John Maynard Keynes, was 'Supply creates its own demand' (Keynes 1936, p. 18). Certainly true until the 1980s was supply creates its own distribution – in search of demand.

For most of human history inventory has equalled wealth. In the 1950s, the president of Toyota declared that inventory is waste. Taiichi Ohno said, 'The more inventory a company has, the less likely they will have what they need' (Liker 2004, p. 104). Toyota was a pioneer in what came to be called just-in-time (JIT) manufacturing. To be effective JIT depends on a deep understanding of demand. Making and delivering where and when demand is expressed eliminates 'waste' and increases profits. By focusing on customer needs and wants business can transition from 'selling supply' to 'serving demand'.

The competitive success of Toyota and other Japanese firms in the 1970s and 1980s converted many U.S. and other manufacturers to similar philosophies and practices. In 1983, *Zero Inventories* by Robert Hall articulated the ultimate goal and a workable process for achieving the goal. It was a process aided and abetted by the increasing ability of computer databases to store and analyse customer data.

In 1972 Walmart had 51 stores. By 1980 it had grown to 276 stores. In 1983 Walmart replaced all of its cash registers with computerised point-of-sale systems. In 1987, Walmart installed the largest privately owned satellite system in the United States to facilitate the increasing amount of demand and sales data being generated. In 1990, Walmart overtook Sears to become the United States' largest retailer. In 1993, with nearly 2000 stores, Walmart achieved its first billion dollar sales week. Today Walmart has nearly 10,000 stores and generates annual sales of over $473 billion. In terms of operating revenue it is the world's largest private enterprise.

There are many explanations for Walmart's phenomenal success, but according to a study by the University of San Francisco, 'Walmart owes its transition from regional retailer to global powerhouse largely to changes in and effective management of its supply chain'. The same study reports:

> Technology plays a key role in Walmart's supply chain, serving as the foundation of their supply chain. Walmart has the largest information technology infrastructure of any private company in the world. Its state-of-the-art technology and network design allow Walmart to accurately forecast demand, track and predict inventory levels, create highly efficient transportation routes, and manage customer relationships and service response logistics. (University of San Francisco 2000)

According to JIT guru Richard Schonberger this technological capability is a very tangible expression of an even more important competitve advantage. Schonberger writes, 'Wal-Mart is the world's grand champion of lean supply chains. While advanced IT gets most of the credit, collaboration is the foundation. Wal-Mart's 2000-odd suppliers near the retailer's Bentonville, Ark., headquarters maintain multifunctional teams on site' (Schonberger 2009). Streamlining and upgrading sources of supply has been fundamental to driving down costs across a wide array of supply chains. This was at the core of restructuring the American supply chain in the last two decades of the twentieth century. An example from the early 1980s,

> Unisys, of New Jersey, cut its supplier base from 750 to 106 and reduced the number of trucking companies from 120 to 8. As a result, purchase costs have dropped by 40 percent, freight costs have been reduced from $.32 per pound to $.09 per pound, and transit times from the Far East have been cut by two-thirds. (Schorr 1992, p. 98)

Success stories like this, and increased competition by early-moving innovators, transformed many product categories from supply–push to demand–pull. In 2000, the Department of Defense opened the global positioning system (GPS) to public use. This transformed the potential of trucks, in particular, to respond to irregular demand as effectively as was once only possible for most favoured customers. Demand increasingly determined supply. In a period of barely three decades, technological innovations have radically transformed 5000 years of transportation history.

The results have also included increasing concentration, greater interdependence and lower profit margins (even as gross profits have increased due to much higher volume). The supply chain at large – and individual supply chains – increasingly involves a complex set of inputs, outputs and multi-layered relationships. And as with other complex entities, the supply chain can sometimes seem (perhaps more than seem) to take on a life of its own.

In his examination of strategies for a catastrophic world Ted G. Lewis explains,

> Complex systems evolve discontinuously through space and time... Such systems appear to have no memory; that is the past is not a prelude to the future. Instead the future of a complex system is highly irregular and unpredictable... Systems become more complex as they are improved; as they are made more efficient, less expensive, and more capable. They also become more self-organized. Therefore the more we improve these systems, the more likely they are to collapse unexpectedly. (Lewis 2012, pp. 341–342)

This is our reality. A pioneering generation of supply chain innovators has succeeded in a historic transition of their field. A rising generation takes the revolution for granted and can barely imagine the pre-revolutionary reality. The successors are naturally inclined to perceive the progress of the last 30 years as predicting the next 30 years. This is not guaranteed. Given persistent patterns of self-organised criticality found in other complex systems, supply chain problems – even potentially catastrophic problems – are likely to emerge.

8.2 COMPLEXITY DRIVES RESILIENCE

Before the technological revolution of the last 30 years the supply chain featured much more redundancy, safety inventory and many more independent players than today. It was less complex, more resilient, more costly and much less efficient.

Resilience is an innate tendency, usually consisting of several inter-related parts, that allows a system to flex under stress and bounce back to something similar to its preexisting condition once the stress is lessened or removed.

Complex systems are not inherently less resilient than non-complex systems. But the resilient characteristics of complex systems usually emerge from recurring experience with cascading – occasionally catastrophic – failures. Failure re-opens a complex system to innovation, adaptation and further optimisation. Can we cultivate a resilient supply chain that avoids – or at least mitigates – catastrophic failure?

The contemporary supply chain continues to self-organise and optimise so rapidly that it is difficult to make any certain claims regarding its innate resilience or non-resilience. But there have been examples – most dramatically after the 2011 Japan earthquake and tsunami – that suggest the global supply chain behaves in a manner analogous to other complex systems.

Looking at other complex systems, three sources of resilience may be especially promising for the supply chain:

- Diversity (especially in terms of roles and functions)
- Decentralisation (and self-organisation)
- Adaptability (even improvisational and opportunistic)

These three characteristics are closely related, especially in terms of the resilient behaviour they support.

Diversity is an effective defence: The functional diversity of a system increases the chance for diversity of response under stress. If even a few individuals or subgroups can effectively adapt to the stress, the entire system is more likely to be preserved and has a chance for recovery. The contemporary supply chain has seen a proliferation of niche players each with very specific functions that contribute to the supply chain's overall health. While there are fewer competitors in any particular niche (less structural redundancy) there has been a significant increase in niches (more functional diversity). This functional diversity is a potentially important source of resilience.

Complex systems are innately self-organising: Feedback mechanisms across a system facilitate the emergence of global patterns from numerous random interactions among widely distributed components. No one controls global effects. Global effects are spawned by spontaneous behaviours that nonetheless produce patterns and rough boundaries that can be predicted. Especially in human-related complex systems dynamic communications among the participants produce shared behaviours to which the system is attracted and around which system equilibrium unfolds.

Over the last 30 years as information has begun to drive the supply chain as much as production or transportation, it has become more and more self-organising. Randomly distributed demand information determines what patterns will emerge in the global supply chain. While supply capacity has become more concentrated in fewer players and places, the number of demand signals has exploded and the number of distribution players remains highly decentralised.

In 1988, Walmart opened its first grocery-embedded 'Supercenter'. There are now over 3200 Walmart locations offering a full line of food and related products. This has transformed the grocery industry and caused particular stress for traditional supermarkets, such as Kroger, Safeway and Supervalu. But the first Whole Foods Market opened in 1980 and in 1988 Whole Foods – playing an upscale niche for 'natural' and prepared foods – began sustained expansion at the same time that Walmart was devouring large segments of the grocery market. Whole Foods now has over 350 locations and generates over $12.9 billion in annual revenue.

In the 1980s, the McLane Company grew from a modest regional distributor to a major national provider of food and other products especially to convenience stores and restaurants. In 1990, McLane was purchased by Walmart. But it was not a good strategic fit and in 2003 McLane was sold to Berkshire-Hathaway and reemerged as a largely independent enterprise. It now generates over $45 billion in annual revenue distributing food mostly to non-grocery stores, the most rapidly expanding segment of the grocery industry.

Walmart is certainly the 'apex predator' in the ecology of groceries. But its disruptive behaviour has opened new niches for other players. Under stress the decentralised, self-organising character of the U.S. grocery market has arguably produced greater differentiation and functional diversity. This is a resilient response.

Resilience can mitigate the negative consequences of change through adaptation: Resilience expects change and spawns structural and behavioural characteristics that accept considerable change as a way of avoiding catastrophe. Diversity does not ensure successful adaptation. A decentralised and self-organising system produces many maladaptive features. But the more diverse and self-organising the system the *more likely* the system will generate – nurture and facilitate – effective adaptation.

In the 1970s, FedEx emerged from a deregulating transportation sector to challenge many long-time participants in the air freight business. In just a few years the veterans experienced a catastrophe. But from the perspective of the full ecosystem – the U.S. economy and eventually the global economy – FedEx was a positive adaptation to changing conditions. The emergence of FedEx, and other supply chain 'species' descended from the FedEx adaptation strengthened the U.S. supply chain and national economy especially by increasing diversity and self-organisation across the supply chain.

Diversity and self-organisation are inputs that increase the likelihood of positive adaptation: Effective adaptation to change – while conserving and strengthening most attributes of the system – is the benefit of resilience. But if adaptation reduces diversity and self-organisation the adaptation is non-resilient and moves the system closer to cascading catastrophe.

8.3   HUMAN RELATIONSHIPS: OPTIMISING DIVERSITY AND SELF-ORGANISATION

Early in the new century a senior executive with a major supplier to Ford Motor Company told researchers from the University of Michigan and Arizona State University, 'In my opinion, [Ford] seems to send its people to 'hate school' so that they learn how to hate suppliers. The company is extremely confrontational. After dealing with Ford, I decided not to buy its cars' (Liker and Choi 2004, p. 106).

In 2010 the North American OEM-Supplier Working Relations survey found, 'For the first time in the history of the 10-year-old study that looks at the relationship between automakers and their suppliers, a U.S. automaker – Ford – ranked in the top three. It was third, with Honda and Toyota in 1st and 2nd place, respectively' (Henke 2010, p. 1). The change in Ford's behaviour has several origins and reflects a broader trend.

In the 1980s, when JIT was initially adopted by many U.S. firms there was much more attention to quantitative outcomes – such as a steep reduction in number of suppliers – than to qualitative inputs, such as the relationships between buyer and supplier. Yet the Japanese practice of JIT emerged from and depended on very dense relationships.

A Japanese *keiretsu* – literally 'headless combine' – is a grouping of firms that collaborate in design, sourcing, production and distribution. They are often linked through mutual ownership, joint planning and a vast web of informal connections. The structural aspects of *keiretsu* have been long recognised by emulators in the United States, but the cultural dynamics of these relationships were not well understood and tended to be undervalued.

There continue to be substantive qualitative differences. In 2010, 21% of Ford's suppliers reported the automaker used threats and retaliation to get price concessions. Only 8% of Toyota's suppliers reported similar behaviour. The Japanese practice of JIT still gives more emphasis to long-term relationships delivering mutual benefit (voice), while the U.S. practice continues to lean towards survival-of-the-fittest (exit). But there is evidence of adaptation across the supply chain eco-system. In mid-2005, John Paul MacDuffie and Susan Helper argued:

> … the 'exit' vs. 'voice' distinction is no longer as clear as it was just twenty years ago. On the 'voice' side, the closed keiretsu system of suppliers characteristic of Japanese industry has been considerably opened to market pressures, requiring more formalization and cost justification of the relationships. On the other, the hard-nosed 'exit' approach of U.S. firms has faced pressure for increased collaboration to achieve the increased levels of quality demanded in the market. There has been a wide range of responses to these pressures, often mixed and contradictory. In the U.S. there are frequent attempts to

achieve the necessary levels of collaboration without trust; but this approach is marked by internal contradictions which, we believe, make it unlikely that it can stabilize as a lasting model. Thus, we will argue, the industry is converging from all sides on a form of pragmatic collaboration, involving substantial levels of trust, though more open and formalized than the traditional Japanese system. (MacDuffie and Helper 2005, p. 429)

This convergence could enhance diversity and self-organisation across supply chains. The classic Japanese model nurtured diversity but suppressed self-organisation. The U.S. approach to implementing JIT has too often ended up with carefully controlled – and risk-increasing – sole sources of supply. While far from reflecting mainstream current practice, there is an increasing realisation that trusted relationships – both tactical and strategic – are essential aspects of an effective and resilient supply chain. Trusted relationships optimise diversity and self-organisation.

To apply for the Malcolm Baldridge Award, sponsored by the National Institute for Standards and Technology, organisations are asked,

What are your key types of suppliers, partners, and collaborators? What role do these suppliers, partners, and collaborators play in the production and delivery of your key products and customer support services? What are your key mechanisms for communicating with suppliers, partners, and collaborators? What role, if any, do these organizations play in implementing innovations in your organization? What are your key supply-chain requirements?

Today it is difficult for many supply chain participants to answer these questions. But there is a growing consensus that these are practically important questions. Being able to honestly respond to these questions with answers that are strategically predisposed to diversity and self-organisation may help mitigate catastrophic consequences.

Complexity cannot be 'managed', as in predicted and controlled. No amount of diversity, self-organisation or adaptation will avoid catastrophe. Complex systems fail. Perturbation, disruption and punctuation are inevitable. The larger and more connected the system, the more likely catastrophe. But it is absolutely possible to anticipate these risks and cultivate a more resilient system.

8.4 MITIGATION NOT PREVENTION

The transformation of supply and demand outlined above has fundamentally altered the supply chain. As recently as the 1980s a chain with links was a reasonable metaphor. That mental image can now dangerously mislead. Rather than a chain it is a stream, a web, a convolution. The system of supply and demand is a network: a complex interlocking set of relationships.

Rather than a linear set of relationships – with an occasional weak link – multiple linkages combine into nodes of varied density. Most nodal connections depend on multiple other connections – across many more nodes – to continue operating even while the entire system is in ongoing flux.

As the supply and demand system has become more networked, it has taken on the core characteristics of other networks. These characteristics include speed, efficiency and flexibility; they also include interdependency, criticality and unpredictability.

The interdependencies are often obscure, usually unrecognised and difficult to accurately identify as long as the network continues moving and morphing. There is an increasing bias towards specialisation which tends to proliferate the total number of suppliers even while it reduces the number of possible sources for any particular specialised component. Suddenly a single small part is unavailable and the whole system shudders.

To be complex is to embrace a totality (Latin: *complecti*, later *complexus*) of many interacting parts. The more complicated the system becomes the more precise organisation is required for the various sub-systems to operate in a predictable way. Paradoxically these organisational improvements can (some argue, *must*) increase the unpredictability of the system-as-a-whole.

Specialisation is often a rational response to accelerating complication. Specialisation can allow an enterprise to perform at very high levels of quality and efficiency. If this quality and efficiency can also be combined with a price advantage, the specialised firm will increasingly attract business from other less specialised and higher cost competitors. Increased volumes allow the enterprise to offer lower per-unit pricing while still producing sustainable revenues even on paper-thin profit margins.

In 2002, two of Japan's largest chip makers, Mitsubishi Electric Corporation and Hitachi Ltd., merged their chip operations into a new $7 billion semiconductor company to be called Renesas Technology Corporation. The founding companies said the move would address short development cycles increasingly required for several types of semiconductor chips (Clarke 2002). A specialised and autonomous company would, it was thought, be able to accelerate decision making.

Early results were promising. By 2007, the company had increased its global market share from less than 4% to more than 18%. As cell phone sales exploded, Renesas was able to ride the wave more effectively than several of its competitors.

The 2007 release of the Apple iPhone began to eat away at market positions for several of the biggest customers for Renesas. Partly in response, in early 2008 Renesas, Sharp Corporation and Powerchip Semiconductors created a joint venture focused on design, development, sales and marketing of LCD drivers and controllers (Powerchip 2008). Renasas eventually became a principal Apple supplier. Another effort to adapt involved accelerated investment in serving the automobile industry with microcontrollers.

In April 2010, the semiconductor operations of NEC were merged with Renesas, creating the fourth largest semiconductor manufacturer in the world. The consolidation was conceived to more effectively deal with increasing competition from South Korean and other chipmakers. Deeper specialisation combined with larger volumes can generate significant advantages in terms of both price and productive capacity. By 2011, Renesas was the global leader in development and delivery of automotive microcontrollers, constituting about 40% of world supply and up to 70% of supply to the Japanese domestic car industry.

On 11 March 2011 a significant earthquake and tsunami struck northeastern Japan. Seven of 22 Renasas production facilities were damaged and closed. Several other corporate facilities outside the hardest-hit area reduced operations due to scheduled electrical blackouts being used to avoid collapse of the electrical grid (Renasas, 14 March 2011). Within 10 days after the initial disaster several of these facilities were able to reopen. But the production facility at Naka – source of roughly one-fifth of Renesas automotive controllers – was especially hard hit. On 28 March, the company announced they did not expect Naka to resume operations until late May (Renasas 28 March 2011). On 11 May, the company announced, 'Those products that will be produced at the Naka factory are expected to gradually ship to customers starting at the end of August. Renesas Electronics expects it will be capable of restoring supply to that of pre-earthquake levels by the end of October' (Renasas 11 May 2011).

The loss of this capacity cascaded through a large portion of the automotive supply chain. Renesas attempted to increase the capacity of other production facilities. Competitors increased production. Automobile manufacturers scrambled to secure alternative sources of supply. In late March *Automotive News* reported, 'Damage from the 11 March earthquake and tsunami at least temporarily knocked out production of 17 of the top 20 Japan-made nameplates sold in the United States, and idled GM and Peugeot factories in the United States and Europe' (Colias 2011). For all of 2011 automotive production in Japan plummeted 12.8% (Organisation Internationale des Constructeurs d'Automobiles 2012). While no automotive assembly facility suffered significant direct damage, the loss of the node at Naka exposed and unwound a critical dependency on which many other nodes relied.

Concentration can enhance efficiency. It can also increase risk. By 2012 Renesas was battling the threat of bankruptcy and as a result of emergency loans had become mostly owned by the Japanese government. In early 2014 the company was still experiencing significant financial problems.

In every prospective particular the consequences of the Naka node's collapse were unpredictable. But while beyond precise prediction, it was also a classic example of how periodic punctuations are a persistent characteristic of large, densely connected networks.

In their seminal 1995 paper, Per Bak and Maya Paczuski argued,

> Large dynamical systems naturally evolve, or self-organize, into a highly interactive, critical state where a minor perturbation may lead to events, called avalanches, of all sizes. The system exhibits punctuated equilibrium behavior, where periods of stasis are interrupted by intermittent bursts of activity. Since these systems are noisy, the actual events cannot be predicted; however the statistical distribution of these events is predictable. Thus, if the tape of history were to be rerun, with slightly different random noise, the resulting outcome would be completely different. Some large catastrophic events would be avoided, but other would inevitably occur. No 'quick-fix' solution can stabilize the system and prevent fluctuations... Large, catastrophic events occur as a consequence of the same dynamics that produces small, ordinary events. (Bak and Paczuski 1995, p. 6691)

Since this insight was suggested, self-organised criticality (or SOC) has been empirically demonstrated to characterise a wide range of biological, physical and social systems. In 1995, the supply chain as a whole was not sufficiently well connected and interactive to demonstrate SOC. But over the last 20 years – as the Renesas example suggests – the increased specialisation, scale and dense connectivity of many global supply chains has resulted in an increasingly efficient network of links and nodes that – precisely as a consequence of these efficiencies – is characterised by SOC.

In other longer-established systems SOC generates infrequent but inevitable catastrophic cascades. In the Renesas example we can see how the interaction of one source of SOC – the movement of tectonic plates – can interact with another source of SOC – the automotive supply chain – in prospectively unpredictable ways. It is reasonable to assume some of these same factors characterise other supply chains (supply networks): water, food, pharmaceuticals, medical goods, fuel, etc. The perverse paradox being: the more efficient the network becomes, the more likely an eventual catastrophic collapse. Risk can be transferred, but it cannot be avoided. As a result, resilience emerges as the best way to 'manage' catastrophic risk.

8.5   ROLE FOR PUBLIC POLICY

A mix of technology and public policy produced the contemporary supply chain, unleashing its creative potential, expanding its complexity, and increasing the likelihood of its catastrophic collapse. While this genie cannot be jammed back in its bottle, can public policy contribute positively to a more resilient complex adaptive system?

The application of network theory, complex adaptive systems theory and SOC to the supply chain is not universally accepted. It is not proven to the satisfaction of many. But recent examples of supply chain behaviour under stress have increased concerns regarding the potential for catastrophic consequences. One such example is the partial collapse of operations at UPS Worldport just before Christmas 2013. UPS is clearly one of the best, most efficient and most effective supply chain operators on the planet. While some externalities, especially weather, contributed to the late holiday disruption, it appears the biggest challenge emerged from unpredicted (unpredictable?) shifts in demand, in other words the supply chain itself was the source of the crisis. What is certainly proven is the role supply chains play in supporting the survival and recovery of communities and regions following a significant natural calamity.

The supply and demand systems and key supply networks have come to be recognised as fundamental elements of economic capacity, national security and disaster management. The supply and demand system is by-and-large a creature of the private sector, but also an important component of national policy.

In January 2012, the President of the United States released a first-time *National Strategy for Global Supply Chain Security*. This initiated a year-long process of private–public consultations. From these discussions a troublesome pattern was identified: for most of the last generation the private sector search for comparative advantage has resulted in a substantial decrease in the supply chain's structural diversity. Fewer firms play increasingly important roles.

Until recently, this structural concentration in the supply chain was not widely recognised by public policy makers. But as it has become more apparent – and its vulnerabilities better understood – the inclination by many in the public sector is to increase regulation. Or as more than one official said, 'We need to require more redundancy in the supply chain'.

The decline of structural diversity in the supply chain is a potentially significant problem that amplifies every threat by reducing the likelihood of innovative responses to stress. But an increase in traditional modes of regulation will – whatever the effect on diversity – reduce the ability of the system to self-organise, undermining the likelihood of innovation under stress or otherwise. This private–public tug of war threatens to wring out of the system two key components of resilience. This is precisely the kind of optimising that contributes to eventual catastrophic failure.

This is no longer a supply chain that can be yanked one way or another. It is much more a spider-web – even more an ecosystem – where what is done in one corner will often have dramatic and unpredictable impacts across the entire system.

Preserving a system's ability to innovate is fundamental to preserving the system's overall integrity. Failures will happen. Innovating around failure is how a complex adaptive system continues to emerge. But long-term health is determined by the sort of innovations and adaptations adopted. When diversity and self-organisation are maximised, resilience is nurtured. When diversity and self-organisation are reduced the next failure is likely to be even worse.

Bureaucracies, both public and private, tend to be suspicious of innovation. There is an inclination to respond to problems with predictable procedures. As procedures accumulate small failures are suppressed and in many complex adaptive systems the likelihood of catastrophic failure is increased. It does not matter if the procedural suppression of innovation is imposed by government regulation or internal management, non-resilience is the outcome.

The *Implementation Update* for the *National Strategy for Global Supply Chain Security* (February 2013) signals a continuing coordination role by White House National Security Staff and a private–public supply chain working group staffed and hosted by one of the cabinet departments. This could be helpful. It might be a waste of time. Over time it could become dangerous. Much depends on behaviour and that is a reflection of who and how and why.

The presumption when private and public meet is – whatever the stated purpose – the eventual agenda relates to rule making and boundary setting *by the government*. As such, the process is subversive to diversity, self-organisation, innovation and resilience. 'Evidence has accumulated that externally imposed rules tend to "crowd out" endogenous cooperative behaviour' (Ostom 2000, p. 147). Stakeholder cooperation in norm setting is especially important in dynamic systems such as the supply chain. In complex adaptive systems rules and boundaries quickly lose influence without active self-monitoring and sanctioning by those involved day-in and day-out, even minute-by-minute within the system. Some studies have found that externally focused regulation can even 'undermine subsequent cooperation' in developing norms (Frohlich and Oppenheimer 1996).

Put another way: When dealing with a complex adaptive system, rule breaking can be sanctioned after the fact by external rule makers. But if prevention of rule-breaking is the goal the rules must be developed and enforced by system participants.

The global supply chain has become a complex adaptive system. As with most complex adaptive systems – and especially human-influenced systems – it is self-optimising. If and when optimisation reduces diversity and self-organisation the likelihood increases for non-resilient consequences and catastrophic failure. Most evidence suggests this is the current trend-line for the global supply chain.

Government action focusing on traditional approaches to regulation will accelerate the movement towards catastrophe. But there is another option. Social manifestations of complex adaptive systems can be influenced by intentional social behaviour. Precisely because government is not engaged as a competitor or vendor in the supply chain (and *is* an important customer), the government might potentially serve as a facilitator or honest broker of boundary setting, rule-making and system sanctioning *by supply chain stakeholders themselves.*

In this vein the June 2014 Quadrennial Homeland Security Review (QHSR), a congressionally mandated strategy document of the U.S. Department of Homeland Security, gives particular priority to how to facilitate the flow of goods in times of crisis,

> Two guiding principles help public-private partnerships maximize the investment by each partner and the success of the partnership: (1) aligning interests and (2) identifying shared outcomes.
>
> By focusing on how interests align, we can provide alternatives to costly incentives or regulations and help ensure a partnership is based on a solid foundation of mutual interest and benefit. There are many examples of public and private sector interests aligning in homeland security. Common interests include the safety and security of people and property, the protection of sensitive information, effective risk management, the development of new technology, reputation enhancement, and improved business processes. New ways of thinking about corporate social responsibility – in which societal issues are held to be core business interests rather than traditional philanthropy – also present an opportunity to identify shared interests.
>
> Where interests do not directly align, potential partners can often be motivated by shared desired outcomes, such as enhanced resilience; effective disaster response and recovery; and greater certainty in emerging domains...

Aligning interests and framing shared outcomes is classic brokering. The modern term broker emerged from the medieval French term for those who broach a wine cask. Who can constructively broach (and assess?) shared issues of threat, vulnerability and risk across the whole supply chain? Is this a role well suited to a public–private panel convened by the government?

The answer depends on who is on the panel and how the principal government officers behave.

The Cross-Sector Supply Chain Working Group called for in the *Implementation Plan* for the *National Strategy for Global Supply Chain Security* could become neutral ground for ongoing communication among supply chain stakeholders. The Working Group would be mostly private sector. If the private sector participants represent actual operational and strategic decision makers, meaningful alignment of interests or outcomes may be possible. The National Security Council official(s) assigned to coordinate this process would certainly not be any kind of White House czar. In most ways s/he would need to be an *anti-czar*, a latter-day Metternich composing and facilitating a supply chain world symphony with as much skill as the old count conducted the Concert of Europe.

This is unlikely. Count Metternich was unique and the complexity of post-Napoleonic Europe pales in comparison with the modern supply chain. Catastrophe is much more likely and from the collapse of market leaders, nation–states and legacy systems something better may emerge. This is the role of catastrophe.

But even if supply chain catastrophe is only to be mitigated, resilience must still be cultivated and this means diversity and self-organisation must be optimised. This combination – diversity, self-organisation, leading to innovation and adaptation – is the strange attractor at the heart of the supply chain's revolutionary transformation. The deeper the system's diversity and the more inclusive its self-organisation the greater the system's potential resilience. Effective mitigation will also benefit from – and probably require – private–public collaboration. A mitigation process could still benefit from the Working Group's engagement, if the engagement is appropriately conceived, scoped and scaled to optimise diversity and self-organisation and thereby innovation.

Can the anti-czar remember this? Can a private–public panel behave consistently with this purpose and direction? Can a complex adaptive system fail and innovate and fail some more, skating along the cusp of catastrophe but be – by application of human purpose – inclined to slide back into the deep basin of diversity, self-organisation and innovation?

Back to Metternich: in his early text, *A World Restored*, a young Henry Kissinger explains how the Austrian count and others crafted the Concert of Europe,

> …the spirit of policy and that of bureaucracy are diametrically opposed. The essence of policy is its contingency; its success depends on the correctness of an estimate which is in part conjectural. The essence of bureaucracy is its quest for safety; its success is calculability. Profound policy thrives on perpetual creation, on a constant redefinition of goals. Good administration thrives on routine, the definition of relationships which can survive mediocrity. Policy involves an adjustment of risks; administration an avoidance of deviation. Policy justifies itself by the relationship of its measures and its sense of proportion; administration by the rationality of each action in terms of a given goal. (Kissinger 1954, p. 326)

If supply chain resilience is to be achieved it must remain a matter of policy rather than administration. Supply chain resilience will not be achieved bureaucratically, but it can be crafted through the intelligent self-conscious give-and-take of authentic private–public collaboration in policy-making. Bureaucratic behaviour – whether it originates in corporate or government offices – threatens the supply chain. Non-bureaucratic policy makers who persistently nourish diversity, self-organisation, and innovation will enable the modern supply chain – and all its benefits – to flourish.

REFERENCES

Bak, P. and Paczuski, M., 1995, Complexity, contingency, and criticality, *Proceedings of the National Academy of Sciences*, 92, 6689–6696.

Bureau of Economic Analysis, 2012, *National Income and Products Accounts*, Department of Commerce, Washington, DC.

Clarke, P., 2002, *Mitsubishi, Hitachi to Merge Chip Businesses*, EE Times, San Francisco, CA.

Colias, M., 2011, Crisis Ends Talk of Price War, *Automotive News*, http://www.autonews.com/article/20110328/RETAIL07/303289956/crisis-ends-talk-of-auto-price-war#.

Frohlich, N. and Oppenheimer, J.A., 1996, Ethical problems when moving to markets: Gaining efficiency while keeping an eye on distributive justice. In: Kaynak, E., Lewis, A. and Arieh, U. (Eds.), *Privatization and Entrepreneurship*, International Business Press, New York, p. 180.

Gilligan, A., 2011, Japan earthquake: Calm after the storm, *The Daily Telegraph, London*, http://www.telegraph.co.uk/news/worldnews/asia/japan/8393056/Japan-earthquake-calm-after-the-storm.html.

Henke, J.W., 2010, North American OEM-supplier working relations study, Planning Perspectives.

Keynes, J.M., 1936, *General Theory of Employment, Interest, and Money*, Palgrave Macmillan, London, Chapter 2, p. 18.

Kissinger, H., 1954, *A World Restored*, Houghton Mifflin, Boston, MA.

Lewis, T.G., 2012, *Bak's Sand Pile: Strategies for Catastrophic World*, Agile Press, Williams, CA, pp. 341–342.

Liker, J.K., 2004, *The Toyota Way: Management Principles and Fieldbook*, McGraw-Hill, New York, p. 104.

Liker, J.K. and Choi, T.Y., 2004, Building supplier relationships, *Harvard Business Review*, 82(12): 104–113.

MacDuffie, J.P. and Helper, S., 2006, Collaboration in supply chains, with and without trust, in Heckscher, C. and Adler, P., eds. *The Corporation as a Collaborative Community*, Oxford: Oxford University Press.

Organisation Internationale des Constructeurs d'Automobiles, 2012, 2011 *Production Statistics*, Paris, http://www.oica.net/category/production-statistics/2011-statistics/.

Ostom, E., 2000, Collective action and the evolution of social norms, *Journal of Economic Perspectives*, 14(3): 147.

Powerchip Corporation, 2008, *Renesas, Sharp and Powerchip to Establish Joint Venture Specializing in Drivers and Controllers*, Tokyo, http://www.psc.com.tw/english/investor/investorEventsContent.jsp?ID=00593.

Renasas Corporation:

http://www.renesas.com/press/notices/notice20110314.html

http://www.renesas.com/press/notices/notice20110328b.html

http://www.renesas.com/press/news/2011/news20110511.jsp.

Schonberger, R., 2009, *The Skinny on Lean Management*, SuperFactory, Online.

Schorr, J.E., 1992, *Purchasing in the 21st Century*, Oliver Wight Companies, New London, NH, p. 98.

Tacitus, Publius Cornelius, at 100 AD, Annals 12.43.2, Rome.

University of San Francisco, 2000, *Walmart: Keys to Successful Supply Chain Management*, http://www.usanfranonline.com/wal-mart-successful-supply-chain-management.

Wright, C.W., 1941, *Economic History of the United States*, McGraw-Hill, New York.

9 Designing Both Systems and Systems of Systems to Exhibit Resilience

C.E. Siemieniuch, M.A. Sinclair, M.J. de C. Henshaw and E.-M. Hubbard

CONTENTS

9.1 INTRODUCTION

In the world of the twenty-first century, just about everything is going to be changing, and this change will be quite fast. According to the latest reports of the UN Intergovernmental Panel on Climate Change, we will be facing more extreme climatic events more frequently, all our systems which make use of energy and on which we rely will be changing, and there are likely to be problems with food and water security (IPCC5-Synth-Rep 2014). This is not necessarily a benign environment; unexpected climate events will happen (heat waves, floods, severe gales, etc.), our transport systems will become electric, as will our homes, and the manufacturing systems that deliver most of our daily needs may change the most, accompanied by occasional disruptions not expected when the changes were planned.

This is a scenario in which governments, organisations and the community need to be resilient, to cope with expected, planned changes happening in parallel, and to cope with the 'oops' scenarios when unexpected events take us by surprise, usually causing some combination of extra costs, delays, damage and sometimes death. It is this latter 'oops' case, likely to become more and more prevalent, with which this chapter is concerned.

By 'coping', we imply that people, organisations and governments will discover the unwanted event, mobilise resources firstly to stabilise the event and prevent it from getting worse, and then restore operations and processes back to what could be recognised as 'normal' behaviour, albeit perhaps a bit different because of any lasting effects of the 'oops' scenario.

If an unexpected event has happened before, it is likely that a plan will exist to get things back to 'normal' again, and this plan will identify what extra resources will be needed (most times, extra resources will be needed), and how to deploy them. If there is no plan, we then have to rely on human ingenuity and experience to restore 'normality' again; to work out what is needed with what resources are available and can be obtained, and then set about restoring the situation.

It is this that is the focus of this chapter – how to be resilient and get back to a satisfactory working state.

The structure of this chapter is as follows. First, we discuss some definitions, both to ensure that readers and the authors have a common frame of reference and to provide some boundaries on the scope of the chapter. We then discuss a number of examples where resilience has been demonstrated and draw lessons from these. The chapter then goes on to discuss how these lessons can be embedded within an organisation; a generic process to embed them, and then the organisational requirements for this process to be successful. Throughout the chapter, the emphasis is on people; this is because they are always at the heart of any recovery effort to bring things back to 'normal'.

9.2   SOME DEFINITIONS

We all have a notion of what resilience means, but it is actually an omnibus concept and has overlaps with other concepts as well, so establishing some clarity early on is advisable. However, given the title of the chapter, we start by defining systems and systems of systems (SoS), and then move on to concepts around resilience.

9.2.1   System

There are many definitions of a system; a useful definition for this chapter is from 'MIL-STD 499B Systems Engineering', a military standard published by the U.S. Department of Defence in 1994:

An integrated composite of people, products, and processes that provide a capability to satisfy a stated need or objective (Appendix A, p. 6).

This definition has the merit of bringing together three classes of entities: people, devices and processes in reaching some goal. However, the definition does not include the notion of a system life cycle, from the conception of a system, through its design, its implementations, its operations and eventually its disposal. This is unfortunate, because through its life cycle (which can last from seconds to aeons) the relationships between these three classes may change, and the components themselves may be replaced. Secondly, for most systems there is an implicit property of hierarchy; typically, one of the set of entities will be in overall control, sending instructions to the other entities and receiving performance feedback from them. Examples are a central heating system in a house, a railway engine, or the GPS satellite system in space.

9.2.2 System of Systems

Jamshidi (2009) has reviewed more than seven potential *definitions of SoS and, although not all are universally accepted by the community, the* following has received substantial attention:

A SoS is an integration of a finite number of constituent systems which are independent and operable, and which are networked together for a period of time to achieve a certain higher goal. (Jamshidi 2009, p. 2)

SoS form a class within the general concept of a system as defined above. The most widely accepted definition comes from Maier (1998):

The elements of the system are themselves sufficiently complex to be considered systems.
Operating together the systems produce functions and fulfil purposes not produced or fulfilled by the elements alone.
The elements possess operational independence. Each element fulfils useful purposes whether or not connected to the assemblage. If disconnected the element continues to fulfil useful purposes.
The elements possess managerial independence. Each element is managed, at least in part, for its own purposes rather than the purposes of the collective.

It is the last point which has the most significance and which distinguishes the class of SoS. Managerial independence implies that the SoS moves towards its goal by negotiation and agreement among the component systems and their owners, not by command. Secondly, any organisation owning one of the component systems can (in theory at least) re-engineer that system to do things differently at its own discretion; as described above, each component system has its own life cycle, apart from the SoS life cycle, which may be much longer. As a result, SoS tend to work by interoperation rather than by integration.

These SoS are very common, and have been around in our societies ever since we started bartering goods and services amongst ourselves. A good example is the SoS that gets you from your front door to a holiday destination far away; buses, trains, airlines and taxies interoperate to accomplish this.

9.2.3 RESILIENCE

A concise definition comes from Collins (2012):

> Recovery to a state that is fit for purpose.

This definition needs a little explanation. 'Recovery' implies that the system or SoS has received a shock that has disturbed its equilibrium and functioning. For example that may mean an external event such as a flood, or a competitor introducing a new product that disrupts the market, or it may be an internal event in which a sub-system, or component system, suffers a significant failure. 'Recovery' also implies that the system, or SoS, as a whole continues to exist (see Section 9.2.4), and that it now mobilises its available resources to restore itself to a state that is 'fit for purpose'.

This last phrase, 'fit for purpose', does not necessarily mean that the system or SoS restores itself to its *status quo ante*. Because the event that caused the shock may be permanent (e.g. a change in government regulations) and the system or SoS may need to reach a different state to meet its new, redefined purpose as a result of the event.

There are some other implications of resilience, too. As Collins said of a system being resilient, 'but how quickly, at what cost, how sustainably, and with what exported effects?' Resilience does mean that the system (or SoS) survives, but it implies additional expenditures of time, effort, money and resources. These aspects are discussed later in the chapter, and elsewhere in this book.

It is important to mention the Risk Management Standard, ISO 31000:2009 (Risk Management Standard 2009) at this point, together with its accompanying guide, *ISO Guide* 73:2009 and ISO/IEC 31010:2009 dealing with risk assessment. This standard, adopted widely around the world, has become a way of standardising the treatment of risk, to be able to assess the vulnerability of companies from a business perspective. For this reason, the terminology of risk has been standardised, as has a generic, high-level process for the management of risk. In this standard, risk is defined as the 'effect of uncertainty on objectives', and therefore implies that risk can have both positive effects as well as negative effects on the achievement of objectives.

These documents set a framework into which resilience fits; resilience provides the conceptual rationale for managing a risk that has been realised, whether it is a performance-enhancing risk or a performance-degrading risk.

9.2.4 SURVIVABILITY

There are considerable overlaps for resilience and survivability, so much so that the main difference is in origin; survivability was developed in the software engineering domain, whereas resilience has a more general foundation.

A number of definitions exist for survivability (Knight and Sullivan 2000, Tarvainen 2004); a working definition, generalising from these, follows:

> Given that a disruptive event happens to a system or system of systems, sufficient of the eight capability assets: finance, knowledge, hardware, software, people, organisation, data/information and logistics are still functional and available and can still be combined to deliver at least a stated acceptable minimum level of service.

The overlaps with the resilience concept are evident. This definition of survivability implies that meta-level resources are available firstly to recognise the nature and scale of the disruption, including what parts of the system(s) are no longer capable (situation awareness), and then to plan and then deliver the restoration of capability, using the eight capability assets, to provide some acceptable level of service. It is obvious that the most important of the meta-level resources is knowledge – the understanding of the system(s), their behaviours, the availability of resources together with their potential for restoration, the understanding and prediction of actual and possible side-effects, and so on. This knowledge is human centred; while humans will be assisted in their planning and restoration efforts by software and hardware tools and techniques, by communication networks and so on, it is the humans involved who are the driving force and decision makers in this restoration effort. Supporting these humans with education and skills training, together with the means to exercise and realise their decisions, is fundamental to successful restoration of capability and, consequently, services. This is particularly important for unexpected, and hitherto unknown disruptive events; in these circumstances it is human ingenuity, expertise and resourcefulness that drives the resilience effort, and all of these qualities are dependent on training, education and experience of the actual, real behaviours of the system and its components. It will also be noted that the experience aspect, given that systems may be distributed geographically, implies that the necessary knowledge for resilience will also be distributed geographically, with an equivalent implication that a communications/knowledge/co-ordinator network entity will be necessary.

9.2.5 AGILITY

As with resilience and survivability, agility has many definitions. A widely accepted definition comes from Alberts (2011, p. 66):

> Agility is the capability to cope successfully with changes in circumstances.

As with resilience, this definition implies that there are resources available that enable the system (or SoS) to be agile. But agility also includes the notion that the system can react fast enough to counteract the event that has caused the need to be agile. Consider an example from the world of supermarkets – a truck delivering containers of milk to a supermarket has broken down. A resilient response will find another truck and supply of milk containers to supply the supermarket, a garage to fix the truck and a way to get the spare milk containers to another market. An agile response will get the containers to the supermarket before the shelves are empty and find another market before the spare containers pass their 'best before' date.

It follows that these two concepts are closely related; a system cannot be agile unless it is resilient, and a system cannot be resilient unless it is agile.

EXAMPLE BOX 1: ORGANISATIONAL RESILIENCE AND AGILITY

1997 – A Japanese company, Aisin Seiki, was a supplier to Toyota, making P-valves, components for brake systems. This company supplied 99% of these components across all Toyota models. Factory No. 1 caught fire; 506 machine tools were destroyed. An alternative supplier was Nishin Kogyo, making the remaining 1%, and unable to ramp up production fast enough to replace the 99%. Toyota at this time was running at 115% of normal production, as a commercial response to impending legislation.

Toyota had only a few hours' stock of P-valves, with trucks on the road carrying another 2 days' capacity. Aisin Seiki salvaged some tools, replaced others and was back in production in 2 weeks, making 10% of requirements and 60% after 6 weeks, and 99% after 2 months.

Aisin Seiki and Toyota each participated in a 'keiretsu' (commercial alliance) and asked for help. 22 from the Aisin Seiki keiretsu and 36 from the Toyota keiretsu replied. Within 5 days, Aisin had made available blueprints and process expertise, and production of P-valves had been allocated. Notably, Denso, a major Toyota supplier, outsourced its own production to free up tools and processes to produce these P-valves (as did others), and helped develop alternative processes for other smaller suppliers. Within 2 days, some P-valves were delivered by these alternate suppliers; within 9 days of the fire, all Toyota plants were functioning as normal again.

During this period, neither financial nor legal negotiations took place, nor was pressure applied to prioritise Toyota. Later, both Aisin Seiki and Toyota recompensed all the companies who had helped in the recovery.

Consider a real-life example of this in the automotive industry, condensed from Sheffi (2005), in 'Example Box 1.'

9.2.6 ROBUSTNESS

Again, this is an omnibus concept. Ross et al. (2008, p. 251), from the perspective of software engineering, have defined robustness as

> Robustness is the ability to remain 'constant' in parameters in spite of system internal and external changes.

This is a definition which can be used in other domains, too; however, it does raise the question of which parameters are we measuring, over what period are we measuring these parameters, and what is our sampling rate? Considering the Toyota case in Example Box 1, if our sampling rate is annual and we assess the supply chain performance parameters just before the fire and then a year later, then the system is undoubtedly robust, because all the resilience activities were agile enough to have restored the parameter values within the year. If we measure on a monthly basis, then a deviance from the parameters will show up, but the values will be restored later on, again showing over the longer term that the system is robust. But if we include internal architecture parameters, then because Aisin Seiki brought in new machines, necessitating changes to processes, we would show that the system was not robust, and is not robust now.

Despite this discussion of the interpretation of parameters, we can all agree that if the parameters show no variation beyond normal as events occur, then our system is robust.

9.2.7 RELIABILITY

There are two interpretations of this term. The first is reliability-as-consistency. In this interpretation, if system parameters are the same at two different time periods, then the system is reliable. However, this interpretation is very similar to robustness, so to avoid confusion, in this chapter this interpretation is not used.

The second interpretation, used in this chapter, is reliability-in-reaching-the-goal-state. In other words, if the system reaches the parameters that define any acceptable goal state (and can do this consistently), then it is reliable.

9.2.8 SYSTEMIC RISK

Using a definition for the 'Global Risks 2014' report (WEF'14 2014), this refers to breakdowns in an entire system, as opposed to breakdowns in individual parts and components. As is shown by the Toyota case study in Example Box 1, a failure in a system component (the fire in Factory 1) can bring an entire system to a halt for a short period; systemic risks, on the other hand, are more pervasive throughout the system, with multiple failings.

Systemic risks as defined in the report are characterised by the following:

- Modest tipping points combining indirectly to produce large failures
- Risk-sharing or contagion, as one loss triggers a chain of others
- 'Hysteresis' or systems being unable to recover equilibrium in time after a shock

These characteristics indicate that restoring a system after the occurrence of a systemic risk may best be characterised as a 'wicked problem' (Rittel and Webber 1973, Daw 2007, Siemieniuch and Sinclair 2014); since the occurrence is likely to be a considerable surprise, then when coupled with the diffuse causation of the surprise and the likely urgency of restitution, some care will be required to avoid making the situation worse; this is outlined in the references cited above.

9.2.9 SOCIETAL RISK

This term is introduced to enable a discussion of the 'levels' of resilience that may be required, with reference to the societies in which we live. The levels of risk (albeit non-unidimensional) are as follows:

1. Individual (e.g. a personal accident)
2. Team (e.g. a team member becomes very ill)
3. Group (e.g. a small village suffers a flood)
4. Organisational (e.g. a fire in a factory)
5. City (e.g. an electrical power outage)

6. Regional (e.g. a severe crop failure; a tsunami)
7. Global (e.g. climate change)
8. Existential (e.g. nuclear war)

Reference to these levels will be made throughout this chapter, to aid clarity. As a further help for clarity, some more example boxes will be found below.

9.3   LESSONS FOR RESILIENCE FROM EXPERIENCE

Resilience has always been a characteristic of the human species and in modern times the lessons are becoming acute. Consider some examples; Box 2, condensed from Amin (2008), describes the Muzaffarabad earthquake in Pakistan in 2005, a natural disaster at level 6 on the societal risk scale. Box 3, condensed from Sanchez-Garcis and Nunez-Zavala, describes an industrial organisational response to a supply chain problem, at level 4, similar to the Aisin Seiki/Toyota example in Box 1 but with significant differences in outcomes (Sanchez-Garcia and Nunez-Zavala 2010). Box 4, condensed from Collins, describes an example of a systemic risk to infrastructure in a city suburb, of limited duration, at level 5.

EXAMPLE BOX 2: EARTHQUAKE

On the morning of 8 October 2005, a massive earthquake measuring 7.6 on the Richter scale struck Afghanistan, India and Pakistan. The earthquake devastated a large portion of Pakistan-administered Kashmir and the eastern districts of Pakistan's North West Frontier Province. About 73,000 people died in the earthquake; 70,000 people were seriously injured and nearly 3.5 million people were rendered homeless. Although this death toll was a quarter of that for the Asian tsunamis, the number of people left homeless was three and a half times greater (ERRA 2006; UNHCR 2006).

The total area affected was 30,000 km², encompassing 9 districts, 25 tehsils (municipalities) and 4000 villages. The damage to economic assets and infrastructure was equally debilitating: 600,000 houses were levelled, and 6298 schools and 796 health facilities were either destroyed or severely damaged (ERRA 2006a). 6500 km of roads were damaged as well. Aftershocks, adding to the damage, numbered 147 the next day and were over 1000 in total.

The initial damage and needs assessment, conducted jointly by the World Bank and the Asian Development Bank between 24 October and 5 November 2005, estimated that the overall cost of recovery would be US$5.2 billion.

A number of practical lessons emerged from this disaster and the subsequent recovery for governments, aid agencies and societies as a whole. These are listed below:

1. Many people refused to leave their land to relocate to more safe and secure areas; this may be partly a cultural phenomenon, and may also reflect the nature of government and state of governance in remote, rural regions. The consequence is that the provision of aid becomes more difficult and dispersed.

2. The initial relief role was provided by the Pakistan Army (mainly transport), assisted by an experienced UN disaster team. They used a 'cluster' approach for identified humanitarian problem domains – food, water, shelter, etc. This arrangement worked well for the initial relief phase, but the reconstruction phase required a more complex approach. Reconstruction sectors were identified: housing, education, health care, livelihood, transportation, agriculture and livestock, environment, power generation, protection, water supply and sanitation, industries and tourism, transitional relief, telecommunications and governance. Guidelines, requiring compliance by reconstruction agencies, were provided for each of these by government departments.

3. Central government has an important role to play in the receipt of aid from external donors: firstly, the role is to ensure that the aid is appropriate to the disaster; secondly that the aid covers all the sectors above that need aid; thirdly that the aid will arrive physically in large airplanes, restricted to long runways, and therefore there must be an organised system for its distribution; and fourthly that the aid arrives in a sequence appropriate to the changing characteristics of the disaster.

4. The handover from relief agencies to rebuilding agencies means new personnel are involved, and many relief personnel leave. There is a handover issue; databases are essential for this, but they are not enough; an 'overlapping' approach is necessary, involving direct human contact to transfer local cultural and political understanding, to build trust and trustworthy relationships and to understand the local geography.

5. Bureaucracy was necessary to ensure good governance and compliance (i.e., answering the questions; 'Are we doing the right things?'; 'Are we doing those things right?' and 'How do we know this?'), but it should not dominate the reconstruction processes. The control and auditing of the disbursement of money from pledges, commitments and national government agencies is an important function.

6. There is an immediate requirement for communication and information resources, especially maps, down to village level. Initially, this was met by a private company (Halcrow Group plc, specialising in infrastructure development; this illustrates the principle that resilience is not solely a government responsibility; it requires both public and private agencies), later fulfilled by RisePak (Research and Information Systems for Earthquakes – Pakistan, developed in conjunction with the UN and the World Bank). As a general comment, digital maps of most of the world exist at a 1:100,000 scale; sufficient to locate most villages and hamlets. Some coverage exists at a 1:25,000 scale. It is likely most developed countries have very detailed maps from satellites, but their availability might be restricted. Latterly, an open-source initiative, 'Missing Maps', has begun to fix this omission.

7. Village-level, even individual-level data and communication resources (such as mobile phones, SMS, crisis-mapping, social media, radios) – are necessary to increase the capacity for affected communities, diaspora groups and ordinary citizens to access, communicate and disseminate useful and

actionable information and to understand the full scale of the relief and restitution effort required and to channel resources effectively.

8. It is important to maintain a local involvement in relief and in reconstruction; the victims of a disaster are also competent people, knowledgeable about local capabilities and able to form relief and reconstruction teams, once they are given trustworthy information and some recourse to aid. Also, the local teams and capabilities remain; they become embedded in local society. The central government has an important co-ordination and governance role, but detailing and execution of plans should be local, entraining local knowledge. The independence of the information from political pressures and private agendas is important for its perceived truthfulness and hence trust, and therefore for appropriate action on the basis of this information.

9. The immediacy of information is important, both for its relevance and for its availability. As one observer expressed it, 'you have about 3 days before the bad guys realise the good guys are no longer in control'.

10. Communication channels have to be able to use social media and old media (e.g. fax). This is because local teams may not have sophisticated facilities, and disruptive events may break normal communications networks. Note that this may preclude the distribution of spreadsheets, maps, etc. However, the global spread of smart phones will become an important communications tool; for example, there are applications to enable direct communications between these smartphones in a 'mesh' network (Simonite 2013); another application (IFRC(TERA) 2010), managed by the International Red Cross and Red Crescent enables it to both broadcast and narrowcast information; also to receive feedback; and power for the phones could be provided by devices such as 'Sunlite™' devices (Lane 2014).

These lessons from a major disaster can be useful for other, lower-level crises where resilience is important. Some of these are universal; for example, that victims are still competent people, and they are there in the crisis; that immediate, high-quality, trustworthy information is an absolute necessity to engage these people as well as to mitigate the spread of fear, uncertainty and doubt; that relief for a crisis is a widespread but local issue; and that the role of government is to entrain private capabilities, to organise the distribution of aid and to provide governance.

An organisational example follows, condensed from Sanchez-Garcia and Nunez-Zavala (2010). This example is at level 4 and is included to show that resilience displayed by one organisation can have serious effects on its competitors; in other words, being resilient may not always be to everyone's benefit.

EXAMPLE BOX 3: NOKIA & ERICSSON

On Friday, 17 March 2000, lightning struck a Philips computer chip fabrication plant in Albuquerque, USA, causing a fire in the facility's area. Less than 10 min later the fire had been put out, leaving firemen with nothing to do when they arrived.

However, smoke damage caused the destruction of millions of chips essential to cellular phones. But the worst damage was to the sterilised fabrication rooms, and it took weeks to restore these rooms to certified, 'clean' status.

On Monday, 20 March, Philips notified its customers of the accident, including Nokia and Ericsson. They accounted for 40% of the affected orders in the Albuquerque plant. When the news reached higher levels in Nokia, it was decided to call this a major incident. The components involved were critical to the company because five different types of chips manufactured in Albuquerque went into Nokia phones.

Two weeks later Philips phoned to explain that restoring the special rooms and resuming production would take several weeks, even months. Nokia now faced a difficult situation because a new generation of cell phones required chips manufactured by Philips. This meant that more than 5% of annual production was liable to be disrupted, with serious potential effects on overall profit margins.

Nokia created a task force with specialised engineers, chip designers and senior Nokia managers to help deal with the problem. As a major client of Philips, Nokia exerted tremendous pressure; they demanded that for a limited period of time Nokia and Philips would work as a single company so as to manufacture the components and get good results quickly. All Philips' spare manufacturing capacity was allocated to meet the needs of Nokia.

Ericsson, a major global competitor for Nokia, also bought a large number of chips from Philips. Ericsson had received the same telephone call from Philips as Nokia the Monday following the fire. Ericsson's reaction, however, was very different; they treated the phone call from Philips on 20 March simply as a business conversation and assumed that since the fire had not inflicted serious damage, Philips would supply the chips a week after notifying the delay, and that everything would be fine. Moreover, Ericsson's middle management did not inform their superiors about likely consequences, even though this implied deferred production.

It was too late when Ericsson finally realised the scope of the problem. They requested assistance from Philips, and Philips informed them that it had allocated all spare manufacturing capacity to Nokia. Ericsson then turned to other component manufacturers but did not find sufficient available options and had to delay production of their new, competing phones.

As a result of the blaze at the Albuquerque plant, Philips lost US$40 million worth of high-technology chips and damage to the plant amounted to €39 million. Nokia suffered minimal losses. However, by the end of 2000, Ericsson posted a loss of US$2.34 billion in the cellular phone division. In April 2001 Ericsson signed a joint-venture agreement with Sony to create handsets. Nokia increased its market share from 27% to 30%, whereas Ericsson fell from 12% to 9%, and in 2011 Ericsson made an exit from the market.

Some lessons from this are listed below:

1. 'Doveryai, no proveryai': 'Trust, but verify' is a Russian proverb; this summarises what Nokia did initially, but Ericsson only carried out the first part. In more technical language, this refers to situation awareness (SA), defined as 'the continuous perception of environmental elements with respect to time and/or space, the comprehension of their meaning, and the projection of their status after a change of some sort occurs' (Endsley 1998, p. 11); an example can be found in Siemieniuch and Sinclair (2004).

In this definition, the two key words are continuity and meaning; Nokia complied, Ericsson less so.

2. Risk management (which is an extension of situation awareness) is of fundamental importance to any organisation; this becomes especially important when there is reliance on one or two sources. Three international standards summarise the approach to risk management: ISO 31000:2009, *Risk Management – Principles and Guidelines*, provides principles, a framework and a process for managing risk. It can be used by any organisation regardless of its size, activity or sector. ISO Guide 73:2009, *Risk Management – Vocabulary* complements ISO 31000 by providing a collection of terms and definitions relating to the management of risk, and ISO/IEC 31010:2009, *Risk Management – Risk Assessment Techniques focuses* on assessment concepts, processes and the selection of risk assessment techniques. The point here is that by carrying out these activities, producing plans to mitigate any of the risks should they occur, and identifying resources to execute the plans, the organisation has gone a long way towards becoming resilient.

3. Speed of response – agility – is the final requirement for resilient response. Nokia demonstrated this and as a result showed a negligible effect in their accounts. Ericsson, by being slow in both their situation awareness and in response to their critical situation, eventually had to leave the market.

4. Resilience may require recourse to other, outside actors; both the Toyota example in Box 1 and this example show the need to find resources outside the normal arrangements. Awareness of the whole market, not just one's own arrangements, is important. The corollary is true as well; unless there is the availability of external resources, reaching and maintaining a resilient status could prove to be expensive.

5. Resilience brings ethical questions as well. It is open to debate whether Philips made the right response to Nokia when the latter demanded that all spare capacity be allocated to Nokia; Philips had other clients, likely to be badly affected by this decision, as the fate of Ericsson demonstrates. It may be that Philips had to respond as it did, because of the terms of the contracts between Philips and Nokia. Whatever, this does indicate that ethics and contract transparency are a part of risk management.

The next example, condensed from Collins, concerns emergent behaviour in complex systems (Collins 2012). This example is at level 5, occurring within a city, and with several lessons.

EXAMPLE BOX 4: SYSTEM OF SYSTEMS BEHAVIOUR WITHIN A CITY

As reported by the BBC: http://news.bbc.co.uk/1/hi/england/london/8431654.stm [In December, 2009], Many residents in north London who had their gas cut off have also lost their electricity after heaters they were given overloaded supplies. About 750 homes in East Barnet [in north London] are still without gas, 3 days after it was cut off. And EDF Energy said about 180 customers have been without electricity since 2030 GMT on Tuesday. A spokeswoman for the National Grid said water from a burst main got into the gas pipes, cutting off the flow. More than 3000 cooking and heating appliances have been distributed to the affected households. And about 150 gas engineers from across the country have gone to Barnet to attempt to get the homes reconnected by Christmas Day. They have already restored supplies to 980 properties. An EDF Energy spokeswoman said: 'Following damage to National Grid's gas mains a number of electric fires have been distributed to residents in East Barnet who were left without heating'. 'The significantly higher demand on the local electricity network has damaged some of EDF Energy's equipment and interrupted power supplies'. EDF is asking customers to use only essential electricity appliances and switch off any non-essential appliances as National Grid carries out its repairs. It has also 'reconfigured the electricity network to make it more robust to help with the extra demand placed on it by the significant increase in the use of electric heaters'.

Most of the lessons arising from this incident concern both legacy and communication issues. East Barnet gradually became absorbed by London as a suburb in the early 1900s, and as the infrastructures of water, waste, electricity and gas were provided as the borough grew, they were buried in pipes under ground along the roads of the borough. This meant that they were in fairly close proximity to each other, creating opportunities for interaction and interference between them. Added to this, all the pipework was put in place by separate agencies, with little or no concern for what other agencies had done, or were about to do. Consequently, due to local geography and soil characteristics, in some places supply pipes could be close to, or even touching, other supplies. It is not unknown for water firstly to excavate the ground underneath the roads and then to penetrate the gas pipes. It being the Christmas season and with a national general election the following year, it is equally not surprising that this was seen as a minor crisis that needed urgent attention; unfortunately, whoever decided that electric heaters were the appropriate answer and also managed to have them distributed swiftly failed to consider the power demands that overloaded the local power grid and produced an even worse situation, now involving people whose water and gas supplies were not interrupted.

This kind of systemic failure due to spatially contiguous systems (this is enough to be considered a system of systems) was first described by Perrow (1999) and is now well known, through many other examples. As a class, these problems fit into the category of 'wicked problems', with known characteristics but with no easy solutions (Rittel and Webber 1973, Jamshidi 2009, Daw 2007, Henshaw et al. 2013, Siemieniuch and Sinclair 2014). For fundamental infrastructures, some of these characteristics (and accompanying lessons) are as follows:

- Neither the infrastructure nor its environment remains stationary. Consequently, continuous situation awareness is necessary ('keep scanning the horizon, and expect trouble'), allied to formal processes for information conservation.
- When a disruption of the infrastructure happens, there is likely to be a cascade of consequential events, both within the infrastructure system itself (e.g. a single failure in an electricity grid that spreads across a region) and in its environment (e.g. Box 4). Prior holistic risk management is necessary both to foresee the potential risks and to take action to mitigate them.

Trade-space analyses (Marr et al. 2006, Roberts et al. 2009) are necessary to identify cost-effective mitigation activities, both prior to any infrastructure upgrades and to deal with any failures. In turn, because a fast reaction may be required, simulations and practices may be necessary to test likely procedures.

- Because unexpected events are likely to occur, perhaps in combinations, those people responsible for dealing with these events need to be orchestrators, networkers, and, on occasion, ambassadors. Since different combinations of people may be enrolled in a team to help, perhaps in roles which are different or flexible compared with normal due to the kind of event and its consequences, there is a strong need for trustworthy relationships among possible team members. This is a cultural issue, requiring development over time (Weick 1991, Blomqvist 1997, Whitworth and Moor 2003, Weick and Sutcliffe 2007), and hence it is also an organisational issue.
- The six principles enunciated by the U.K. Royal Academy of Engineering are relevant to many scenarios where engineering is required to restore the required status (RAE-CSTW 2007):
  - Debate, define, revise and pursue the purpose
  - Think holistic
  - Follow a systematic procedure
  - Be creative
  - Take account of the people
  - Manage the project and the relationships

However, when the engineering scenario can be described as 'complex' (and increasingly this will be case as time rolls on and systems become ever more integrated and interoperated), these principles will require augmentation; a good source for this is the *INCOSE Capability Systems Engineering Guide* (Kemp and Daw 2014).

9.4   SOME GENERIC LESSONS EXTRACTED FROM THESE EXAMPLES

So far, a number of exemplars and their lessons have been discussed. Distilling these down to a common set across all these exemplars to what might be considered a sparse, generic set gives us the following:

- The exemplars above all show that resilience efforts have people at their core. It is true that these people may be supported by a vast array of technology and technical services, but it is the people that drive the processes. The U.S. Air force long ago developed the 'OODA Loop'; Observe, Orient, Decide, Act, and this decision-making loop is a key to resilience activities. It is because people are the repositories of much formal and tacit knowledge, both foreground and background, and are the only sources of authority and responsibility over resources; this places them at the centre of the effort, executing the OODA loop. In most cases there will be a team involved (because only for very small crises will one person have all the requisite knowledge and authority); since agility is likely to be important, it follows that trust is a fundamental property for such teams. Trust here refers to trust within the team (i.e., that people will be honest about what they can and cannot do, and that when people promise to do something, they will deliver on that promise. It also refers to trust between people and the technology and technical resources that they employ (Blomqvist 1997, Whitworth and Moor 2003).

- Recovering from a complex, disruptive, and likely distributed event may have the characteristics of a 'wicked' problem (Rittel and Webber 1973, Daw 2007, Siemieniuch and Sinclair 2014). Two characteristics of wicked problems are that firstly, the class of solution that restores 'fitness for purpose' in the definition of resilience quoted above, depends on the initial starting point; the recognition that an event either has occurred or is about to occur, and the identification of this event. This requires continuous, holistic situation awareness, perhaps best summarised by a business quotation from the 1990s: 'Be afraid', attributed to the INTEL™ CEO, Andy Grove. Secondly, an appropriate, resilient solution may not be evident until it is achieved. This problem may be curbed by the efficient and timely handling and classification of disaggregated data, coupled with a focus on local teams, the training of these teams and the provision of support, all of which will reduce many of the confusions characteristic of wicked problems and will reduce them to a more manageable scope.
- Tracking and maintenance of disaggregated data. Aggregated data is of little use when systems or components are in danger of failing or have failed because aggregation removes the detail in which many little warnings of imminent, or actual, failure will be found. Timely, organised, informative, disaggregated data is what is essential, but it needs information and communication systems for its collection, organisation and presentation for disaggregated data to be available and useful.
- There is a need for local daily assessment of the needs of distributed, localised teams with local knowledge working to deliver their part of a resilient solution, and then to deliver the support that is needed to them. This creates a requirement for a flexible, 'extreme programming' kind of approach to finding a solution that turns out to be 'fit for purpose'; an example is the SCRUM approach, applicable to resilience (Schwaber 2004).
- For local teams to be effective, they have to know *a priori* what to do, where they can call on what types of information and what support they can expect. From a local perspective, the SCRUM approach (or similar) needs to be embedded (Beck and Andres 2004). But for this to work, trained people, in positions of authority and responsibility and with resources to exercise them, are essential.
- Quality control of data and its availability is a fundamental necessity. Note that quality in this case includes timeliness, as well as accuracy, precision and relevance.
- The internet is vital for coordination, but only with appropriate protocols, procedures and information security provision. Furthermore, the

realisation of the Internet of Things (more accurately the Internet of Things, People and Services) may become a significant resource for resilience activities, given that the enterprise may have difficulties in mobilising its own resources quickly enough. Few organisations, in whatever field they operate, can afford to have spare resources always available for emergencies.

- Feedback is essential to people, teams and organisations involved in resilience activities, so open access and reports on status are necessary. Note that this enables transparency, accountability and participation; however, security will then become a significant issue. Feedback is also necessary for any external organisations that become involved in the resilience effort; donors, governments and other capability suppliers will require reassurance that their efforts are being utilised appropriately, and are not being appropriated for other purposes.

We now turn to a more structured discussion of resilience to indicate how these lessons could be incorporated into practice.

9.5   EMBODYING RESILIENCE

As stated earlier, ISO 31000 sets out a high-level process for risk management. Assuming that an organisation has carried out a recent risk assessment exercise in its continuous risk management process and has defined one or more risks, it is now time to be more specific about preparing for each risk.

On the basis that 'a picture is worth 1000 words', Figure 9.1, adapted from Mackley (2008) summarises the sequential and parallel activities that are necessary to be confident that the organisation is resilient with respect to the risk that is being addressed.

The column headers comprise the eight categories that contribute to resilience. It is expected that the columns will be processed in parallel, subject to cross-links and dependencies between them. The row headers provide sequencing for the activities in the columns, again subject to cross-links and dependencies.

The diagram can be understood in three ways. Firstly, 'Are we ready to be resilient?' Assume that a particular risk has occurred today. Addressing this event today can only happen with whatever resources are available today (row 1 in Figure 9.1), and the capability that these resources represent depends on how complete have been the activities carried out the week before, the month before, and so on.

Secondly, 'How prepared do we need to be?' Depending on the results of risk assessment, the management might wish to start from row 5 in Figure 9.1 and work up through the rows to reach a state of readiness that is deemed acceptable, given the risks identified.

Thirdly, there is an agility aspect to this figure, answering the question, 'Now what do we do?' Given that an event has occurred that requires a resilient response, the time available in which to respond determines how resilient we can be. For example, if we have 1 day, then the response will be determined by what is available today. However, if we have a week, or a month, in which to respond (because

		Resilience enablers							
		Finance	Knowledge	People	Organisation	Software	Hardware	Data/Information	Logistics
Preparation time for resilience (also time to plan and assemble more resources to restore capability)	Today	Commence payment for work	Utilise on-site knowledge	Deploy team(s)	Commence restoration of services	Engage software	Deploy hardware	Implement plan	Support teams
	Week	Fund practices	Finalise practices for implementing plan	Assemble for practices	Organise assembly of teams	Test software	Position hardware	Present plans	Support planning and pre-positioning of hardware and people
	Month	Organise financial resources	Maintain training knowledge for hazards, disruptive events	Continue training of individuals and teams	Maintain organisational readiness	Maintain software readiness	Maintain hardware readiness	Maintain plans for hazards, disruptive events	Support maintenance activities
	Year	Envisage financial requirements for hazards, disruptive events	Organise training knowledge for envisaged training needs	Train people in responding to hazards, disruptive events	Maintain teams for responding to hazards, disruptive events	Ensure appropriate software is available and maintained	Ensure that hardware is fit for purpose and is maintained	Utilise data and metrics to provide situation awareness	Support organisational preparedness
	Several years	Ensure financial stability of the enterprise	Continually assess and address knowledge needs	Recruit and train appropriate people	Organise 'good governance' within the organisation	Prepare appropriate software assets	Prepare appropriate hardware assets	Ensure situation awareness for operations, hazards	Organise appropriate logistics support

FIGURE 9.1 Preparing for resilience and deploying teams to deliver resilience. Cross-links between cells have been omitted for clarity.

e.g. there have been warning signals sensed in our situation awareness systems), it is then possible to entrain some of the resources in rows 2, 3 and lower, to produce a more wide-ranging, encompassing response; perhaps by retraining, by releasing technical resources from current commitments, and so on. This could include the time required to reach outside the organisation to find other available resources; much as Toyota and Aisin Seiki did in Box 1, or as Pakistan did to deal with the aftermath of the earthquake.

At this point, we have discussed what is meant by resilience, what lessons apply, and indicated the processes by which a resilient state could be reached. We now turn to the question of 'How do we organise to get there?'

9.6   ORGANISING FOR RESILIENCE

In the paragraphs above, it was stated that people are at the core of resilience; they provide the goals and the impetus, they execute their responsibilities for resilient behaviour of the enterprise by exercising their authority over available resources, and as necessary, innovate when faced with the unexpected. It was also stated that trust between people, and trust between people and their technology and technical services is fundamental to recovery. The discussion below is organised firstly into the functional nature of trust, and then the organisational characteristics necessary to support trusting, resilient behaviour at the organisational, workgroup and individual levels.

Other good discussions, making many of the same points, will be found in Siemieniuch and Sinclair (1999), Weick et al. (1999), Roberts and Bea (2001), Hollnagel et al. (2006) and Woods (2006).

9.7   THE IMPORTANCE OF TRUST IN RESILIENCE AND RECOVERY

What binds a company together is not its technical quality and expertise, nor is it the professionalism of its management, important though these things are; the real glue is the organisational and human quality of trust. Because of the centrality of humans in organisations, systems and processes, trust between humans is of critical importance. However, as we move forwards into a world of cyber-physical systems and the Internet of Things, where the devices we use and command are likely to be interconnected and have some level of intelligence embedded within them, trust in the capabilities and functioning of these devices becomes more and more important. For this reason, we first discuss trust between humans, and then trust between humans and machines.

9.8   HUMAN–HUMAN TRUST

There are many definitions of trust (Blomqvist 1997, Whitworth and Moor 2003), and for this chapter we use the following definition:

> Trust is the belief in the behavior of others – that they will behave with integrity and will deliver what they promise on time, and in full; and that they will not promise what they cannot deliver, and if circumstances interfere with delivery, these will be explained as fully as possible, including attempts to circumnavigate these circumstances.

Without some degree of common purpose and trust in the behaviour of others, there is likely to be little forward progress. Trust relies on a number of things:

- The establishment of common goals. It does not mean that all the goals of the company have to be known and supported by everybody, only that the important ones should be held in common. This is a strategic issue for senior management.
- Transparency about problems, and ways of working. This includes such notions as open-book accounting, open and measured internal processes,

open acknowledgement of errors and so on. In particular, this implies that mistakes should not be seen as opportunities for enforcing discipline, but treated as opportunities to learn.

- A willingness to share benefits. Resilience implies that deep learning will occur, and it is likely that financial and other benefits will accrue as a result. Arranging for these benefits to be shared equitably among the participants will again require policies to be established and is another strategic issue. The Aisin Seiki/Toyota example in Box 1 is relevant to this.
- A common understanding of terms and their usage. This is important at the management and operational level, emphasised in ISO 31000 on risk management and endemic in the increasing adoption of standards. It is also a matter of training and practice to embed this understanding.
- Respect for confidentiality, both at a professional and a personal level. This is a standard requirement within all organisations, strongly affected by the organisational structure and role design. It is a cornerstone of trust.
- Speedy and efficient execution of promises. This is another of the cornerstones of trust and of effective companies. The important determinants of this are organisational design; empowerment (responsibility, authority and access to resources) of individuals and teams; effective control of processes; and access to timely and relevant knowledge and information.
- Personal relationships, built up over time. This, together with the next point, comprise the other two cornerstones of trust. Knowing your peers' capabilities, empowerment, biases and foibles provides the unspoken context by which any dialogue is translated into meaning. It is the basis on which you can rely on your peers to take the right actions for the effective operation of the company and its processes (Suchman 1987, Mantovani 1996). The quality of these personal relationships is affected very much by the organisation's human resource policies and by its approach to the empowerment of individuals. In particular, peer group networks, a rich source of knowledge for resilience, will not happen successfully without the generation of personal relationships.
- Recognition of the 'favour bank'. As complexity theory indicates (Siemieniuch and Sinclair 2002, Allen et al. 2005, Richardson 2005), the behaviour of an organisation is complex and not fully predictable, and people are always liable to be surprised by some event, which usually has to be resolved in a hurry. At this point one may call on others to provide help, resources, or alternatives. This may be accomplished by individuals going outside normal procedures ('bending the rules') to resolve the situation. In so doing, they 'bank a favour' to be redeemed when they themselves have a similar untoward event. The ability to be able to work at the interstices of organisational design in this unofficial manner is a core competence in any company, and is directly affected by the culture and level of empowerment of the individual person, and is therefore an organisational design issue.

We now move on to consider human–device trust, as a special case of human–human trust.

9.9   HUMAN–DEVICE TRUST

This is an area of current, rapidly expanding research, so the comments below should be seen as tentative.

Most of us have learnt to trust the devices that we use; the electric kettle boils water and switches itself off, the vacuum cleaner sucks and we understand its little foibles, and the car works well until it needs a service. For most devices, the definition of trust given above still fits and human–device trust is engendered.

However, once intelligence is built into devices, and these devices are networked, the situation changes. For very limited intelligence, such as the refrigerator informing us that item X has reached its storage limit, or the vacuum cleaner informing us that it really needs emptying and the filter needs cleaning, there is no great impairment to trust; these machines just carry on behaving as they do, according to our expectations. However, once their intelligence reaches a level where autonomous operation and learning is involved, the situation changes; we have entered the realm in which devices may exhibit behaviour changes. An obvious example is the household robot, which will become more useful to you as it learns your habits and adjusts its behaviour accordingly. But for health, safety and security reasons, it may record your behaviour and transmit it to some analytical entity outside your direct control (e.g., insurance companies, maintenance centres and so on) thereby creating an identity for you over which you have no direct control. And there may be reasons for these companies, or hackers, to reprogram the robot.

Summarising this: devices may not behave as you expect and may utilise information in ways that are opaque to you. Most times, this will be benign, but there may be occasional unpleasant surprises, and it is these that cause lasting mistrust.

This issue will become more and more important as cyber-physical systems (robots being one example of these) become more intelligent, more capable, more networked and more pervasive in our societies, due to their positive effects and benefits.

In a resilience context, where SoS contain cyber-physical systems as components, risk management according to ISO 31000 will have to take these into account. The problem for trust arises when some untoward or unexpected event occurs that affects the cyber-physical systems, perhaps arriving through its network connections. This event may be a failure in an external sensing system, now sending incomplete information, or a malignant attack by hackers, or any of a number of other causes. Two classes of problems emerge; the first is what might be termed the 'SysAdmin' problem, where the resilience/recovery effort is aimed at the compromised cyber-physical system to fix it, while the rest of the system of systems carries out a work-around to continue normal functioning, and the second is where some event has occurred and we now wish to use the cyber-physical system in the recovery effort and discover that it no longer behaves as expected.

Summarising again, most times trust in devices is perfectly acceptable and safe. What damages trust is unexpected changes in behaviour, and the introduction of intelligent, probably networked, devices capable of single-, double- and triple-loop learning increases the chances of this being experienced. We do not yet know how to warn users about such changes in behaviour nor of the likely behavioural effects. If we cannot work this out beforehand, no doubt bitter experience will show us.

9.10   THE ORGANISATIONAL PERSPECTIVE

At the organisational level, the development of a culture of trust also depends on strategic leadership and support. We list some of the important considerations below:

- Leadership. Without this guidance there is a danger that semi-autonomous teams will overdevelop their working cultures, procedures and knowledge in isolation to address the dilemmas of the moment, rather than collaborating to address the longer-term strategic resilience needs of the business. It will also help to diffuse the problem illustrated in the next point below.
- Instantiation of a seamless core IT and communications infrastructure to support knowledge and information dissemination and team coherence and coordination. Clearly, this is a critical resource, as discussed earlier. It is therefore vital that a suitable IT&T infrastructure is in place, with the right applications implemented. There must also be the right information- and knowledge-sharing policies in place too; as a minimum, we believe that it is necessary to allow an individual access to operational information at the managerial level above, and operational level below, that individual's position in the hierarchy. This would be the minimum to ensure that the individual has situation awareness of longer-term issues, emerging consequences, and detailed performance. This also applies to project teams working on different resilience aspects of the developing situation; their personnel must be aware of what is going on as well, since there is a need to present a co-ordinated, cohesive front to the rest of the world.
- Development of processes that will support organisational performance under 'normal' circumstances, and recovery processes for those systems essential to the oganisation's integrity and continued existence.
- Appropriate performance criteria, reward structures and training provision that encourage people to work collaboratively in a trustworthy manner and to develop and utilise new knowledge.
- An official strategy and supporting policies for appropriate devolution of responsibility and authority to a range of system stakeholder groups. This empowerment is of critical importance for resilience; as said earlier, the 'victims' are those with the most localised knowledge and awareness of work-arounds and are likely to be better at predicting local consequences of actions.
- Proper resourcing of processes. This includes sufficient human resources of sufficient quality – over-zealous downsizing undertaken by company strategists is a particular danger (Haigh 1992, Thornhill et al. 1997, Kozlowski and Ilgen 2006). It suffices to point out that downsizing often affects those of middle years, in middle management positions. The consequences of this are that (a) every time a person is removed, there is a net loss in problem solving and innovative capability; (b) the organisation's corporate memory resides largely in the minds of the middle-aged and this is effectively thrown away and (c) there is a loss of corporate morale,

as Haigh (1992) has pointed out. It has escaped few people in industry that one attempt at downsizing is usually followed by another in the near future. Furthermore, the destruction of the informal relationships and communication channels that have been built up and which play a major part in the efficient running of many businesses damages the efficiency and capability of resilience efforts.

- A strategy for the maintenance of resilience. Firstly, this can be interpreted as adopting and implementing ISO 31000 and its supporting standards as a core part of the organisation's business. This will identify risks that may threaten the organisation's existence, for which recovery processes will need to be planned. But planning is not enough; resilience capability will be verified and validated only by simulations, demonstrations and practice, in whatever combination is thought necessary.

9.11   THE WORKGROUP PERSPECTIVE

This is predicated on the proposition that in most organisations the unit of organised human work has become the team, rather than the individual operator, and it is the team that will deliver resilient performance when this is required.

- Provision of organisational structures, roles and rewards that support team-working and job security. In turn, these outcomes of organisational design will foster the development of a culture of openness and information sharing important to resilience. If responsibility and authority are to be vested in teams, then it follows that the people in those teams should be both enabled to execute their responsibilities and to do so efficiently and with good motivation.
- Provision of policies for operation of the workgroup, including group leadership; decision making; problem resolution, etc. These policies all directly influence the effectiveness of information and knowledge utilisation. Furthermore, as a particular issue, policies must cover the mode of operation of the workgroup – for example, as a crew (little or no overlap of skills within the group, and specific people are allocated to specific tasks) or a team (much overlap of skills and knowledge within the group, allowing opportunistic allocation of people to tasks, depending on current conditions), or something in between.
- Devolution of responsibility and authority to the workgroup, summarised as 'sufficient authority to make mistakes and the responsibility to retrieve them' (within given limits). This has been discussed above; the important concomitant of this philosophy is that the business processes involved should be well-defined; otherwise, the mistakes may become disasters rather than 'opportunities to learn'. Note that this learning covers localised knowledge and work-arounds, mentioned earlier, both essential for recovery efforts.
- Provision of appropriate communication channels, and sufficient communications content regarding policy, developments, plans and so on to ensure

coherence, consistency, cohesion, co-ordination, continuity and conformity in actions and decisions. This should include specific consideration of the needs of peer group networks.
- Provision of appropriate technical resources and training (including support from the IT&T infrastructure) to be able to execute the group's processes efficiently on a regular basis and to make use of new knowledge, both for 'normal' operations and for recovery situations.
- Provision of workgroup technical support to enable the application of knowledge. This includes job aids, manuals, software, etc.; those things which are necessary for everyday functioning of the work group and for recovery efforts.

9.12   THE INDIVIDUAL ROLE VIEWPOINT

At this level, the case studies to be found in papers on high reliability organisations are relevant (Bigley and Roberts 2001, Roberts and Bea 2001, Rosness et al. 2001, Weick and Sutcliffe 2007). Summarising the findings from these and other studies leads to some principles as follows:

- People who commit (or are committed) to promises must be empowered to execute them, with appropriate knowledge, skills and authority to use resources in doing so.
- There should be appropriate reward structures, both financial and non-financial, that reflect the knowledge needs of the organisation or system for resilient action.
- Peer group networks transcending the boundaries of organisations must be created to propagate the flow of tacit knowledge and to provide additional sources of expertise and action when resilience is required. It is here that the concept of the 'favour bank' becomes especially important.
- Providing people with a better understanding of their work situation and giving them the confidence to act on their own initiative come from practice; the concept of 'learning by doing, utilising small-scale simulations and exercises based on 'what-if'' scenarios is necessary to provide this understanding. This is an area where book learning is insufficient.
- Education, not just the training of people. Contacts with local colleges and other educational establishments are vital to provide the background education necessary to understand the 'why' that determines the 'how'.

9.13   CONCLUSIONS

This chapter has covered considerable ground, ranging from examples of resilience in action to aspects of how to embed resilience in organisations, down to the level of the people within them. This was deliberate; it is people who are central to any recovery effort that occurs in the name of resilience, and if your people are resilient, then, given the right technical tools, resilience will always be an attribute of your systems.

REFERENCES

Alberts, D. S. 2011. The agility advantage. US DOD Command & Control Research Program.

Allen, P. M., J. Boulton, M. Strathern and J. Baldwin 2005. The implications of complexity for business process and strategy. *Managing Organisational Complexity: Philosophy, Theory and Application*. K. Richardson (ed.). Mansfield, MA: ISCE Publishing, pp. 397–418.

Amin, S. 2008. Data management systems after the earthquake in Pakistan: The lessons of RisePak. *Data against Natural Disasters: Establishing Effective Systems for Relief, Recovery, and Reconstruction*. S. Amin and M. Goldstein (eds.). Washington, DC: World Bank, pp. 233–271.

Beck, K. and C. Andres 2004. *Extreme Programming Explained: Embrace Change*. Boston, MA: Addison-Wesley.

Bigley, G. A. and K. H. Roberts 2001. The incident command system: High reliability organizing for complex and volatile task environments. *Academy of Management Journal* 44(6): 1281–1299.

Blomqvist, K. 1997. The many faces of trust. *Scandinavian Journal of Management* 13(3): 271–286.

Collins, B. 2012. *Building Resilience into Complex Systems*. Loughborough University, UK, 18 June 2012.

Daw, A. J. 2007. Keynote: On the wicked problem of defence acquisition. *7th AIAA Aviation Technology, Integration and Operations Conference: Challenges in Systems Engineering for Advanced Technology Programmes*. Belfast, NI: AIAA, pp. 1–26.

Endsley, M. R. 1998. Situation awareness, automation and decision support: Designing for the future. *CSERIAC Gateway* 9(1): 11–13.

ERRA (Earthquake Reconstruction and Rehabilitation Authority). 2006. *Rebuild, Revive with Dignity and Hope: Annual Review 2005–2006*. Islamabad.

Haigh, G. 1992. *The Fetish for Sacking. The Independent Monthly*. Surrey Hills, Sydney, NSW, pp. 12–15.

Henshaw, M., C. E. Siemieniuch, M. A. Sinclair, S. Henson, V. Barot, M. Jamshidi, D. Delaurentis, C. Ncube, S. L. Lim and H. Dogan 2013. Systems of systems engineering: A research imperative. *IEEE International Conference on System Science and Engineering (ICSSE)*. A. Szakal (ed.). Budapest, HU: IEEE.

Hollnagel, E., D. D. Woods and N. Leveson (eds.) 2006. *Resilience Engineering*. Aldershot, UK: Ashgate Publishing Ltd.

IFRC(TERA) 2010. TERA (Trilogy Emergency Relief Application) and beneficiary communication. *IFRC* http://www.ifrc.org/en/what-we-do/beneficiary-communications/tera/ 2014.

IPCC5-Synth-Rep 2014. *Climate change 2014 synthesis report: Approved summary for policymakers*. UN Intergovernmental Panel on Climate Change, New York.

Jamshidi, M. (ed.) 2009b. *Systems of Systems Engineering – Principles and Applications*. Boca Raton: CRC Press.

Kemp, D. and A. Daw 2014. *INCOSE UK Capability Systems Engineering Guide*. Ilminster, UK: INCOSE UK.

Knight, J. C. and K. J. Sullivan 2000. On the definition of survivability. Department of Computer Science, University of Virginia.

Kozlowski, S. W. J. and D. R. Ilgen 2006. Enhancing the effectiveness of work groups and teams. *Psychological Science in the Public Interest* 7(3): 77–124.

Lane, E. 2014. How technology is changing disaster relief. BBC News.

Mackley, T. 2008. *Concepts of Agility in Network Enabled Capability. Realising Network Enabled Capability*. Oulton Hall, Leeds, UK: BAE Systems.

Maier, M. W. 1998. Architecting principles for systems-of-systems. *Systems Engineering* 1(4): 267–284.

Marr, O. S., M. Waters, G. Tidhar, L. Bache, M. Ling, G. Mathieson and M. Selvestrel 2006. Developing a requisite analytic trade-space for assessing Agile Mission Grouping – Approach adopted for the construction and implementation of the DARNSTORMS model. *11th ICCRTS – Coalition Command and Control in the Networked Era*, Cambridge, UK: US DOD.

Perrow, C. 1999. *Normal Accidents – Living with High-Risk Technologies*. Princeton, NJ: Princeton University Press.

RAE-CSTW 2007. *Creating Systems that Work: Principles of Engineering Systems for the 21st Century*. London: Royal Academy of Engineering.

Richardson, K. A. (ed.) 2005. *Managing Organisational Complexity*. Greenwich, CT, USA: Information Age Publishing.

Risk Management Standard 2009. http://www.iso.org/iso/home/standards/iso31000.htm.

Rittel, H. W. J. and M. M. Webber 1973. Dilemmas in a general theory of planning. *Policy Sciences* 4: 155–169.

Roberts, C. J., M. G. Richards, A. M. Ross, D. H. Rhodes and D. E. Hastings 2009. *Scenario Planning in Dynamic Multi-Attribute Tradespace Exploration. SysCon2009 – IEEE International Systems Conference*. Vancouver, BC: IEEE.

Roberts, K. H. and R. Bea 2001. When systems fail. *Organizational Dynamics* 29(3): 179–191.

Rosness, R., G. Håkonsen, T. Steiro and R. K. Tinmannsvik 2001. *The Vulnerable Robustness of High Reliability Organisations: A Case Study Report from an Offshore Oil Production Platform*. Trondheim, Norway: SINTEF Teknologiledelse, pp. 9.

Ross, A. M., D. H. Rhodes and D. E. Hastings 2008. Defining changeability: Reconciling flexibility, adaptability, scalability, modifiability, and robustness for maintaining system lifecycle value. *Systems Engineering* 11(3): 246–262.

Sanchez-Garcia, J. and M. Nunez-Zavala 2010. Nokia and Ericsson, IPADE. Universidad Panamericana, Aguascalientes, Mexico.

Schwaber, K. 2004. *Agile Project Management with SCRUM*. Redmond, WA, USA: Microsoft Press.

Sheffi, Y. 2005. *The Resilient Enterprise: Overcoming Vulnerability for Competitive Advantage*. Boston, MA: MIT Press.

Siemieniuch, C. E. and M. A. Sinclair 1999. Knowledge lifecycle management along the supply chain. *Human Computer Interaction*. Munich: Erlbaum.

Siemieniuch, C. E. and M. A. Sinclair 2002. On complexity, process ownership and organisational learning in manufacturing organisations, from an ergonomics perspective. *Applied Ergonomics* 33(5): 449–462.

Siemieniuch, C. E. and M. A. Sinclair 2004. Long-cycle, distributed situation awareness and the avoidance of disasters. In *Human Performance, Situation Awareness and Automation: Current Research and Trends*. D. A. Vincenzi, M. Mouloua and P. A. Hancock (eds.). Mahwah, NJ: Lawrence Erlbaum Associates, Vol. 1, pp. 103–106.

Siemieniuch, C. E. and M. A. Sinclair 2014. Extending systems ergonomics thinking to accommodate the socio-technical issues of systems of systems. *Applied Ergonomics* 45(1): 85–98.

Simonite, T. 2013. A crowdfunding campaign to set smartphones free from cellular networks. *MIT Technology Review*. http://www.technologyreview.com/view/517106/a-crowdfunding-campaign-to-set-smartphones-free-from-cellular-networks/2014.

Tarvainen, P. 2004. Survey on the survivability of IT systems. *Proceedings of the 9th Nordic Workshop on Secure IT Systems (NORDSEC 2004)*. Espoo, Finland.

Thornhill, A., M. N. K. Saunders and J. Stead 1997. Downsizing, delayering – but where's the commitment? The development of a diagnostic tool to help manage survivors. *Personnel Review* 26(1): 81–98.

UNHCR (Office of the United Nations High Commissioner for Refugees). 2006. *The State of the World's Refugees 2006*. New York: Oxford University Press.

WEF'14 2014. *Global Risks* 2014 – *Ninth Edition. Insight report*. Geneva: World Economic Forum.

Weick, K. E. 1991. Organisational culture as a source of high reliability. *California Management Review* 29: 112–127.

Weick, K. E. and K. M. Sutcliffe 2007. *Managing the Unexpected: Resilient Performance in an Age of Uncertainty*. San Francisco: Jossey-Bass.

Weick, K. E., K. M. Sutcliffe and D. Obstfeld 1999. Organizing for high reliability: Processes of collective mindfulness. *Research in Organisational Behaviour* 21: 23–81.

Whitworth, B. and A. D. Moor 2003. Legitimate by design: Towards trusted socio-technical systems. *Behaviour and Information Technology* 22(1): 31–51.

Woods, D. D. 2006. Essential characteristics of resilience. *Resilience Engineering*. E. Hollnagel, D. D. Woods and N. Leveson (eds.). Aldershot, UK: Ashgate Publishing Ltd., pp. 21–34.

10 Organisational Resilience and Recovery for Canterbury Organisations after the 4 September 2010 Earthquake

H. Kachali, J.R. Stevenson, Z. Whitman,
T. Hatton, E. Seville, J. Vargo and T. Wilson

CONTENTS

10.1   INTRODUCTION

On 4 September 2010 at 4:35 a.m., the Canterbury region of New Zealand was shaken by a Richter M_w 7.1 earthquake. The epicentre was in Darfield, a town approximately 40 km west of Canterbury's largest city, Christchurch. The intensities of the event in different geographic locations ranged from MM3 (felt but little damage) to MM9 (considerable damage – partial collapses) on the Modified Mercalli scale (USGS, 2009) as illustrated in Figure 10.1. There were no fatalities from the event (as a consequence of the early hour of the morning); however, extensive damage occurred to buildings and infrastructure, with significant liquefaction effect on the soil, in many areas of the city. Organisations in both the rural and urban areas of the region already affected by the impacts of the global recession were now faced with recovery resulting from the earthquake and subsequent aftershocks. The post-disaster environment offered many challenges but also many potential opportunities for organisations.

This chapter presents the results of a survey which forms the first part of an ongoing study investigating the factors influencing recovery for individual organisations and industry sectors in Canterbury following the 4 September earthquake. The survey explored organisational impacts, challenges, mitigation and preparedness in relation to this event for both urban and rural organisations. This chapter is an updated version of an article published in 2012 (Kachali et al., 2012) with the addition of suggestions for practice which incorporate the researcher's reflections, 3 years and many more earthquakes later. Understanding how post-disaster outcomes manifest for different types of organisations and industry sectors will help organisations and

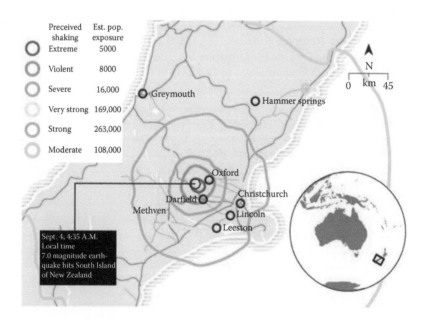

FIGURE 10.1   Darfield earthquake map. (Adapted from http://eqclearinghouse.org/co/ 20100903- christchurch/.)

decision makers tailor planning, mitigation, response and recovery strategies that are appropriate for different sectors and for different geographic locations.

The Canterbury region has a population of 521,832 and is a significant part of the New Zealand economy. Half of all South Island businesses, accounting for 53% of South Island employees, are located in Canterbury (Statistics New Zealand, 2011). Figures from the New Zealand Treasury Department (2010) put the combined loss from the 4 September earthquake at $5 billion NZD. This aggregated amount masks the variations among individual organisations, sectors and geographic locations. In addition, the effects of disaster exceed physical damage to buildings, stock and infrastructure. Organisations affected by disaster face disruptions that flow on to the community and other organisations that depend on them (Tierney and Nigg, 1995; Webb et al., 1999). The stated loss estimates do not include these other significant costs of recovery such as business interruption, decreased customer numbers and property devaluation experienced by organisations post-disaster (Rose and Lim, 2002; Wood, 2008).

Recovery from disaster is a complex and interconnected process and is not a guaranteed outcome for affected organisations. Recovery is defined here as 'longer-term efforts to reconstruct and restore the disaster-stricken area e.g. through repairing or replacing homes, businesses, public works and other structures' (Tierney, 1993, p. 1).

10.2   LITERATURE REVIEW

There has been an increasing trend in the communicated cost of natural disasters globally (Munich Reinsurance Company, 1999) and recovery from these disasters can account for a significant proportion of national economies (Munich Reinsurance Company, 1999; Benson and Clay, 2004). Direct losses include damage to premises, infrastructure, equipment and loss of revenue resulting directly from the event (Cochrane, 2004). Indirect losses, which are difficult to measure, include income loss due to supply chain issues or decreased sales caused by customer income losses (National Research Council, 1999). It has been shown that indirect losses can surpass property damage in cost and pervasiveness (National Research Council, 1999). Rose and Lim (2002) state that business interruption losses are possible even without physical or property damage and can result from interdependencies and flow-on effects between organisations, employees, suppliers and customers.

The trajectory of economic trends within business sectors is influenced by disasters (Benson and Clay, 2003). For instance, it is expected that the retail sector suffers loss of revenue while the construction and manufacturing sectors experience a boom in the wake of a disaster (Boarnet, 1997; Tierney and Webb, 2001). According to Registered Master Builders New Zealand, there was a dip in construction sector revenue after the September 2010 event which only started to rise in February 2011 (RMBF, 2011). Tracking the timing and distribution of economic impacts across sectors is important for providing appropriate support for sectors post-disaster.

In the literature, it is recognised that factors such as the type of organisation and industry sector, the size of the organisation and its location contribute to how different organisations and sectors recover from disaster. Other factors that contribute to recovery include the age of the organisation, owning or renting business premises and the level of organisational disaster preparedness (Tierney and Dahlhamer, 1997;

Alesch et al., 2001). For example, it is thought that smaller organisations may be more vulnerable due to restricted access to resources and networks that are available to larger organisations (Chang et al., 1996; Alesch et al., 2001). A contrary view is that small size may be advantageous as it allows flexibility and ability to adapt rapidly to changed circumstances (Gunasekaran et al., 2011; Vargo and Seville, 2011). It is acknowledged that post-disaster recovery of small business plays a vital role in the economic and social recovery of a community (Pelling, 2003).

Furthermore, there is an emerging body of literature that links an organisation's level of resilience to its recovery (Chang et al., 2001; Bruneau et al., 2003). Resilience is an umbrella concept reflecting an organisation's ability to not only survive but to be able to thrive through times of adversity (Seville et al., 2008). More needs to be understood about how different types of organisations are affected by disaster, the factors that influence recovery, and how long after disaster they recover (Galbraith and Stiles, 2006).

10.3 METHODS

This survey was designed to capture the initial impacts and perceptions of organisations affected by the 4 September earthquake. The survey employed a combination of concepts from qualitative and quantitative research. Data were collected using Dillman's (1978) total design method, adapted to this work. Questionnaires were mailed to organisations. This was followed by a telephone call where organisations were given the option of completing the survey by phone or in a personal visit with a member of the research team, using an online survey tool or returning it by post or e-mail. The multi-media approach was designed to cater to those organisations that might have relocated, closed or were too busy to complete the telephone survey during work hours. The final response rate was greatly improved by the flexible format approach to data collection.

The survey was designed to collect information from organisational leaders/managers about their organisations' experiences following the Darfield earthquake. The authors wanted to capture early assessments from organisations about how they were affected, ways they felt they had mitigated those effects and to identify potential challenges as their recovery continued. Many of the specific impacts organisations were asked about in the survey were derived from the organisational as well as the disaster literature. In addition, the survey contained open-ended questions that asked organisations to detail the effects of the earthquake on their organisation. The survey also included a shortened form of the Organisational Resilience Measurement Tool (McManus, 2008; Lee et al., 2013) developed by the Resilient Organisations Research Programme. This was used to obtain a snapshot of the resilience profile of sampled organisations based on the 13 indicators described in Table 10.1.

10.3.1 SAMPLE

A cross section of industry sectors was strategically selected for this study to reflect various elements of the Canterbury economy. Within each of these sectors, organisations were randomly selected to take part. The sectors included were

TABLE 10.1

Indicators of Organisational Resilience and Items Included in the Survey

Indicator of Organisational Resilience	Item Included in Survey 1 Questionnaire
Leadership	There would be good leadership in our organisation if we were struck by a crisis
Information and knowledge	If key people were unavailable, there are always others who could fill their role
Recovery priorities	Our organisation has clearly defined priorities for what is important during and after a crisis
Decision making	When we need to, our organisation can make tough decisions quickly
External resources	Our organisation keeps in contact with organisations that it might have to work with in a crisis
Situation monitoring and reporting	Our organisation monitors what is happening in its industry to have an early warning of emerging issues
Planning strategies	I believe that the way we plan for the unexpected is appropriate, given the people and organisations that count on us
Proactive posture	Our organisation is focused on being able to respond to the unexpected
Participation in exercises	Our organisation understands that having a plan for emergencies is not enough and that the plan must be practised and tested to be effective
Minimisation of silos	Our organisation works hard to remove barriers for working well with each other and other organisations
Internal resources	During business as usual, we manage resources so that we are able to cope with a small amount of unexpected change
Staff involvement	People in our organisation are committed to working on a problem until it is resolved
Innovation and creativity	People in our organisation are known for their ability to use their knowledge in novel ways

- Information and communication technology (ICT) – a high-growth sector identified as a key component of Canterbury's regional economic plan
- Critical infrastructure (lifelines) – for provision of services vital to recovery
- Hospitality (cafes, restaurants and bars) – to analyse recovery through consumer discretionary spend
- Fast moving consumer goods (FMCG) – including product producers, supermarkets, convenience stores and petrol stations to analyse recovery through consumer non-discretionary spending
- Trucking – important part of supply chain and logistics for many industry sectors
- Building suppliers (wholesale and retail) – for their involvement in the rebuilding process
- Christchurch and Kaiapoi Central Business Districts (CBDs) – because they are retail hubs and represent an aggregation of organisations in one locality
- Rural farm – organisations close to the fault trace and also a high-growth part of Canterbury's regional economic plan
- Rural non-farm – organisations supporting rural communities

10.4   RESULTS

The results of this survey highlight the effects of the 4 September earthquake on the Canterbury economy by analysing impacts to particular sectors and the possible interdependencies between them. In the first part of the survey, organisations were asked for demographic information. Respondents were then asked whether they had been affected by the 4 September earthquake. Those that responded 'no' were directed to complete only the organisational resilience portion of the survey. Eighty percent of sampled organisations reported having been affected by the earthquake. All results herein, which describe organisational impact and mitigation information, are from organisations that reported being 'affected' by the earthquake.

10.4.1   SURVEY RESPONSE RATE

Of the 869 organisations contacted for the survey, 376 usable responses were returned, giving an overall response rate of 36%. The industry sectors with the highest response rates, by percentage, were ICT and critical infrastructure while that with the lowest was rural farm. Figure 10.2 shows the response rates for all the sectors sampled.

10.4.2   ORGANISATION LEVEL INFORMATION

Table 10.2 shows the average number of employees and periods of operation of respondents, by sector. Respondents consist primarily of small businesses reflecting

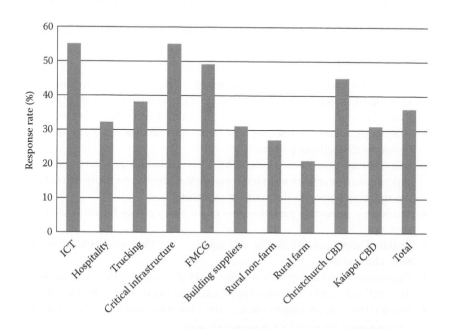

FIGURE 10.2   Organisational resilience and recovery survey – response rate by sector.

TABLE 10.2
Organisation Level Information

| Sector | Number of Employees | | | | | | Years in Operation | |
| | Full-Time | | Part-Time | | Temporary | | | |
	Mean	Median	Mean	Median	Mean	Median	Mean	Median
ICT	18	6	2	2	3	1	15	11
Hospitality	9	5	18	7	1	1	13	10
Trucking	31	10	9	2	1	1	33	24
Critical infrastructure	233	112	41	13	192	4	80	100
FMCG	154	75	63	52	4	0	38	24
Building suppliers	11	7	2	1	1	1	25	20
Rural non-farm	9	2	3	2	11	2	25	11
Rural farm	35	2	2	2	1	1	34	28
Christchurch CBD	15	3	29	3	2	2	35	30
Kaiapoi CBD	5	3	4	2	1	1	35	20
Total	46	5	17	3	26	1	31	19

New Zealand's organisational demographic profile (Statistics New Zealand, 2008). Organisations from the critical infrastructure and FMCG sample groups were found to be the oldest organisations and with the largest number of employees in the sample. At the time of the survey, 67% of respondents had been in operation at least 10 years.

10.4.3   AFFECTED ORGANISATIONS AND DURATION OF CLOSURE

The highest proportion of respondents who reported being affected by the earthquake was from the Kaiapoi and Christchurch CBDs, hospitality and critical infrastructure sample groups. Organisations in the Kaiapoi CBD were affected by extensive liquefaction and lateral spread while those in the Christchurch CBD were likely more affected by the official cordons placed around the CBD in the days and weeks after the earthquake. The high number of organisations affected in the hospitality sector corresponds with their location; a large portion of the hospitality sample was located in or around the Christchurch CBD. Lastly, the proportion of critical infrastructure organisations affected might be due to the placement of their infrastructure making it especially vulnerable to ground shaking (e.g. buried cables or pipes) and services in high demand immediately after the earthquake. The percentage of affected organisations, by sector, is shown in Table 10.3.

Sixty-three percent of affected organisations closed for some time following the earthquake. Rural farm and FMCG organisations closed for the least amount of time. For rural farm, this is due in part to farming organisations being unable to close in the way that organisations in other sectors would. On average, organisations from the trucking sector and Kaiapoi CBD were closed the longest. The average duration of closure for the entire sample was 7 days.

TABLE 10.3
Affected Organisations and Duration of Closure

Sector	Affected by 4 September 2010 Earthquake (%)	Duration of Closure (days)	
		Mean	Median
ICT	56	3	2
Hospitality	94	8	7
Trucking	71	11	2
Critical infrastructure	92	4	3
FMCG	88	2	1
Building suppliers	70	3	2
Rural non-farm	88	5	4
Rural farm	67	2	2
Christchurch CBD	90	9	7
Kaiapoi CBD	90	11	7
Total	80	7	4

Organisations were presented with a list of reasons that may have contributed to the organisation's closure after the earthquake. The two reasons most cited for closure in the CBDs and the hospitality sector were 'building waiting to be structurally assessed' and 'damage to immediate locality'. Approximately 50% of respondents also cited 'clear up damage to interior' as one of the reasons for closure. In addition, closure because of 'stock loss or damage' featured prominently for the FMCG and hospitality sectors. Reasons for this include breakage caused by shaking, loss of refrigeration due to power outages and the short shelf life of perishables.

Only 25% of trucking organisations reported closing for any period of time following the earthquake. This is likely due to locational flexibility (e.g. many can operate to some extent even with limited access to their building) and also because of the minimal earthquake damage to road networks they use.

10.4.4 SUPPLY CHAIN

The closure of some organisations had supply and demand-side effects on other organisations. The trucking industry reported that one of their challenges was the lack of warehousing with widespread collapse of storage racks in warehouses including in two regional food distribution centres. Trucking organisations could not deliver goods because receiving organisations were closed but then faced increased demand for trucking and supply services when organisations were ready to re-stock.

When organisations were asked about the ability of their regular suppliers to meet their needs, 46% of affected organisations reported their regular suppliers were 'completely capable' while 26% reported that they were 'somewhat capable'. For FMCG however, close to 57% of respondents reported their suppliers as being 'somewhat

TABLE 10.4
Capability of Regular Suppliers

	Capability of Regular Suppliers			Need to Use New Suppliers	
	Completely Capable (%)	Somewhat Capable (%)	Completely Incapable (%)	No (%)	Yes (%)
ICT	40	13	2	53	4
Hospitality	47	38	6	78	13
Trucking	42	21	5	63	5
Critical infrastructure	58	33	0	75	17
FMCG	29	57	2	62	26
Building suppliers	53	17	3	70	3
Rural non-farm	55	21	5	74	12
Rural farm	50	13	3	63	3
Christchurch CBD	52	27	6	85	3
Kaiapoi CBD	50	25	3	80	10
Total for entire sample group	46	26	4	69	10

capable' and only 28% thought their suppliers 'completely capable'. Twenty-six percent of FMCG and 17% of critical infrastructure organisations reported the need to use new suppliers. For critical infrastructure, this might be the result of the sharp increase in products used for repair and replacement after the earthquake. Information about supplier capability for each sector is presented in Table 10.4.

10.4.5 INSURANCE

Insurance against hazard events can be used by organisations to mitigate loss. However, deciding the financial value for insurance of low probability events such as earthquakes can be difficult (Kunreuther, 2006). Sampled organisations were presented with a list of insurance options. Across the sample, the most common types of insurance were 'public liability' (60%) and 'organisation assets and equipment' (59%), as seen in Table 10.5.

Sectoral differences in insurance may reflect differing organisational requirements as well as varying perceptions of risk. The two most cited types of insurance for ICT organisations were 'public liability' and 'assets and equipment'. This is likely because ICT organisations want to protect themselves against claims should their clients suffer loss and seek recompense and also because they rely heavily on their equipment for the operation of the organisation. Critical infrastructure organisations were also more likely to have 'public liability' insurance than any other. The ICT and rural farm were the least likely to have 'cash-flow, income protection and organisation interruption' insurance, while the hospitality sector and organisations in the Christchurch CBD sample were most likely to have it. This may reflect the varying nature and importance of cash transactions for the different sectors.

TABLE 10.5

Percentage of Organisations with Different Insurance Types

Sector	Cash Flow, Income Protection and Organisation Interruption (%)	Property and Buildings (%)	Assets and Equipment (%)	Motor Vehicles (%)	Public Liability (%)	Commodities and Goods – Stock (%)	Others (%)
ICT	24	27	49	31	44	20	15
Hospitality	78	47	75	44	69	63	0
Trucking	37	45	50	47	50	24	16
Critical infrastructure	38	54	50	54	67	33	42
FMCG	62	57	62	62	64	60	21
Building suppliers	43	43	57	57	63	50	10
Rural non-farm	43	76	69	57	69	50	2
Rural farm	23	63	33	63	47	33	7
Christchurch CBD	70	48	73	55	64	70	12
Kaiapoi CBD	45	40	68	38	68	35	13
Total	45	49	59	49	60	43	13

In the 'other' insurance category, some of the building suppliers and critical infrastructure organisations reported being self-insured. Self-insurance is often opted when organisational wealth is higher and risk perception is relatively low (Ganderton et al., 2000). In the case of critical infrastructure, following several hazard events in Canterbury in the early 2000s, private insurance costs had increased dramatically, leading to a perception that insuring assets was uneconomical. Instead, some organisations adopted other mitigation measures such as seismically reinforcing structures housing important assets and upgrading equipment to decrease the risk of loss (Eidinger et al., 2011). These decisions contributed positively to a rapid restoration of critical utilities such as power and water networks (Giovinazzi et al., 2011).

Organisations were also asked about their relationships with their insurer, their banker and also how satisfied they were with their insurance package on a scale from 'very dissatisfied' to 'very satisfied'. From the overall sample, 18% of organisations reported feeling 'very satisfied' with their insurer while 29% were 'satisfied'. Twenty-four percent of all sampled organisations were 'very satisfied' with their banker and 25% were 'satisfied'. More organisations from FMCG than from any other sector reported being 'very satisfied' with their insurer, insurance package and banker at 36%, 33% and 38%, respectively.

10.4.6 Decisions Affecting Recovery

Organisations work in an increasingly interdependent environment. Post-disaster decisions made by others, over which organisations have little control, can greatly influence organisational recovery. In our survey, organisations identified several decisions or actions that occurred external to the organisation that negatively influenced their ability to recover, including

- Damage to nearby buildings (other building owners' decisions not to seismically retrofit their buildings)
- Official cordons around nearby buildings
- Delayed insurance payouts
- Duration of ongoing building inspections
- Road closures
- Official curfew (7 p.m. to 7 a.m.) briefly instituted in Christchurch following the earthquakes

Organisations reported all of the above as having flow-on effects that led to varying levels of business interruption and loss of revenue. As a result of building inspection delays, owners could not access their premises, and in some cases employees were reluctant to work from buildings they perceived as unsafe. In addition, some organisations in the Christchurch CBD reported that cordons around nearby buildings gave customers the perception that the CBD was 'closed'. Also, as the rebuilding work did not commence as quickly as some expected after the earthquake, partly due to inspections and delayed insurance payouts, some building supply organisations reported difficulties deciding what material to stock or produce for when the work commenced.

Furthermore, the decisions made by an organisation in the immediate aftermath of disaster can influence not only their long-term recovery but that of other organisations (Dietch and Corey, 2011). In the survey, organisations were presented with the short-form of the Benchmark Resilience Assessment Tool (Whitman et al., 2013). This tool gives a series of statements about the organisation and asks to what extent the respondents agree or disagree with the statements. Results for the total sample showing only 'agree' and 'strongly agree' are given in Table 10.6.

The critical infrastructure sector had the largest numbers of organisations 'strongly agree' with all the statements. This is likely because critical infrastructure organisations realise how vital they are to other organisations and to the community. They are also more likely to engage in preparedness exercises and have a formal response plan. More organisations 'agreed' that 'the way we plan for the unexpected is appropriate, given the people and organisations that count on us' than with any other statement.

Organisations from the building supply and rural farm sectors were more likely to 'agree' that 'there would be good leadership if our organisation were struck by a crisis'. Overall, 54% of organisations 'agreed' with this while 35% 'strongly agreed'. This corresponds with the high number of all organisations (89%) that either 'agreed' or 'strongly agreed' with the statement 'when we need to, our organisations can make tough decisions quickly'. Regardless of industry, sector or organisation size, good leadership is necessary for decision making and to effectively manage staff in a high stress crisis environment.

For the statement 'our organisation has clearly defined priorities for what is important during and after a crisis', 49% of the total sample 'agreed' and 24% 'strongly agreed'. These priorities may be specified as part of the organisation's crisis preparedness activities and adapted as necessary post-disaster. These preparedness activities could include defining the minimum resources the organisation needs to get through a crisis and the steps necessary to ensure staff well-being and business continuity.

For 'our organisation keeps in contact with organisations it might have to work with in a crisis', 50% of organisations from both the FMCG and rural farm sectors 'agreed' and 54% of critical infrastructure 'strongly agreed'. This is in contrast with organisations from building suppliers, ICT and hospitality where only 3%, 5% and 6%, respectively, 'strongly agreed'. Having information on, for example, where and how to access aid or what part of the supply chain is broken can help an organisation's recovery. The highest percentage of organisations to 'agree' that their 'organisation monitors what's happening in its industry' were from the Christchurch CBD (64%), rural farm (60%) and ICT (58%). For rural farm and ICT, this is likely because the trends in these sectors change very often. Also, knowledge of industry trends can be used to formulate corporate strategy post-disaster. For instance, an organisation might diversify to other markets while its local market was in recovery.

The hospitality sector had the highest number of organisations (75%) 'agree' that 'the way we plan for the unexpected is appropriate, given the people and organisations that count on us'. For some, this may not reflect that there is extensive planning but simply that the owners perceive their sector to be non-critical.

TABLE 10.6
Organisation Level Statements

		ICT (%)	Hospitality (%)	Trucking (%)	Critical Infrastructure (%)	FMCG (%)	Building Suppliers (%)	Rural Non-Farm (%)	Rural Farm (%)	Christchurch CBD (%)	Kaiapoi CBD (%)	Total for Entire Sample Group (%)
There would be good leadership in our organisation if we were struck by a crisis	Strongly agree	40	28	45	58	48	17	38	20	30	23	35
	Agree	49	53	45	42	45	77	48	67	64	65	54
If key people were unavailable, there are always others who could fill their role	Strongly agree	15	13	26	46	31	7	21	20	21	18	21
	Agree	55	50	50	46	48	67	45	53	55	45	51
Our organisation has clearly defined priorities for what is important during and after a crisis	Strongly agree	11	22	24	58	36	10	26	20	21	23	24
	Agree	45	53	42	38	48	57	50	60	61	45	49
When we need to, our organisation can make tough decisions quickly	Strongly agree	36	28	50	50	60	23	43	33	42	30	40
	Agree	53	53	34	50	33	60	43	53	58	60	49
Our organisation keeps in contact with organisations that it might have to work with in a crisis	Strongly agree	5	6	21	54	24	3	21	23	21	13	18
	Agree	35	44	37	46	50	27	40	50	42	48	41
Our organisation monitors what is happening in its industry to have an early warning of emerging issues	Strongly agree	24	19	34	63	45	13	38	13	24	25	29
	Agree	58	47	39	33	40	47	48	60	64	50	49
I believe that the way we plan for the unexpected is appropriate, given the people and organisations that count on us	Strongly agree	16	0	32	54	26	7	14	10	15	18	19
	Agree	65	75	39	42	55	70	64	63	67	53	59

10.4.7   CHALLENGES AND OPPORTUNITIES

Disasters present both challenges and silver linings for organisations. In this survey, organisations were asked to report the 'biggest challenges' they faced following the 4 September earthquake. Across all sectors and geographic areas, the most commonly reported 'biggest challenge' was the well-being of staff. However, other sector specific challenges as well as opportunities also emerged.

Apart from difficulty forecasting demand, building suppliers also reported reduced sales while they waited for the rebuilding work to begin. Organisations in the construction industry were aware that there would eventually be a surge in demand for their services, but delays caused by ongoing aftershocks and lags in insurance pay-outs made it difficult to predict when reconstruction work would begin. Further, uncertainty about employment prospects might lead to skilled workers in this industry migrating out of Christchurch, causing a skills shortage when the rebuilding work starts in earnest (Tertiary Education Union [Producer], 2011). Conversely, as the economic landscape of Canterbury has changed, new skills will be required across many sectors, partly to re-train people who lost their jobs to re-enter the job market and also because of the need for specific skills (e.g. insurance loss adjustors) as a result of the earthquake (TVNZ, 2011).

The CBD and hospitality sectors cited cash flow, reduced customer numbers and reduced consumer spending as major challenges. This could be a result of changed consumer habits as they reduce spending due to uncertainty about the future economic climate. It could also be due to consumers continuing to shop in the suburbs even after the CBD shops reopened. The hospitality sector also noted problems with staff availability. This might be due to population outflow after the earthquake or because staff were not prepared to work from the CBD due to the perception that buildings were unsafe.

Several ICT organisations, on the other hand, reported their biggest challenge was dealing with increased demand for their services. This is possibly due to organisations adopting new technologies after the earthquake to conduct their business, as part of hazard mitigation and preparedness, as well as the need to repair and replace damaged equipment. The biggest challenges for trucking, rural non-farm and FMCG included issues with supply chain and logistics.

10.5   CONCLUSIONS: SUGGESTIONS FOR PRACTICE

Post-disaster recovery is a complex economic, political, social and physical process. The physical damage to an organisation's property is only one small part of the lingering disruptive effects following a major event. The results from this study demonstrate the complexities of the increasingly interdependent world with supply and demand issues along with spatial dependencies identified as areas where disruptions occurred. The study also highlights the importance of people including both the ability of leaders to respond appropriately and employees to be able to function in a time of high stress and distraction. The results also illustrate the need to consider carefully insurance arrangements.

10.5.1 INSURANCE

From the entire sample, only 47% of organisations were satisfied or very satisfied with their insurance package. Other on-going research by Resilient Organisations indicates a lack of understanding from many organisations of just what coverage is provided by their policy and of the time frames for claims settlement (Brown et al., 2013). This most particularly applies to business interruption type insurances where the applicability of the coverage was not as was expected in many cases. Organisations need to ensure that care is taken over understanding the policy conditions and making the appropriate trade-offs regarding cost and cover. The decision to invest in mitigation measures, rather than insurance, greatly influenced the rapid restoration of essential services, contributing enormously to the ability of other organisations to recover. There may be lessons from this applicable to other sectors with regard to considering all possible mitigation measures and not just insurance alone.

10.5.2 SUPPLY CHAIN ISSUES

Availability of and access to resources is integral to the recovery of an organisation following disaster. However, disasters can disrupt the web of interactions in a supply chain in ways that are difficult to predict. Results from this survey show that 28% of organisations were effected by suppliers being 'somewhat capable' following the earthquake. More recently, the Thailand floods of 2011 demonstrate the cascading effects on other organisations both regionally and internationally. Souter (2000) writes on the importance of organisations not only managing their own risks but also those of the connections in their supply chain.

Knowledge of suppliers and customers occupying critical positions in an organisation's supply chain is crucial to recovery for two major reasons. First, an organisation affected by disaster can act to mitigate ripple effects to its important supply chain partners. Second, organisations can develop contingency plans in case their core suppliers or customers face disruptions that could affect their operations (Finch, 2004). Also, having a good pre-disaster relationship with organisations on either side of the supply chain can be an effective way of ensuring that your business is viewed favourably when there is intense competition for goods or services. Organisations need to balance the concepts of lean production and just in time manufacturing with the need to withstand some supplier disruption. Additionally, organisations need to consider how well placed they are to capitalise on sudden peaks in demand both in terms of stock or equipment and with regard to human resources.

10.5.3 EXTENT OF DISRUPTION

Disruption was encountered by a large number of organisations in Canterbury which did not necessarily suffer any direct damage. One of the most obvious examples of this was organisations with relatively undamaged premises who were unable to access their property due to a collapse risk from nearby buildings. When examining the risks that your organisation faces, consideration needs to be given to the vulnerability not just of your premises, but also your staff, customers, suppliers, neighbours, local area and critical utilities.

10.5.4   WELL-BEING OF PEOPLE

The survey found the well-being of staff to be the biggest post-earthquake issue for all sectors. Many preparedness actions and plans deal with buildings and equipment but make little allowance for the status of their people. Enduring a major earthquake, can be a significant trauma for individuals regardless of the presence or absence of physical damage in their immediate environment (Walker et al., 2013). Additionally, the sight of damage to their friends' or families' property along with public areas and buildings can cause significant emotion and stress. Beyond the psychological effects, practicalities such as access to water and power may be an issue for some. Dealing with the after effects in terms of disruption to facilities, instigation of insurance claim processes and providing support needed by others may consume much time and energy. Organisational preparedness measures for both disasters and day to day crises need to ensure that there is back-up and support plans in place for people, including senior leaders who may be amongst those most personally affected by a situation.

10.5.5   READINESS FOR OPPORTUNITIES

Disasters result in enormous changes in demand patterns as well as spatial behaviour patterns. Organisations in Canterbury that were able to rapidly adapt and respond to the changes prospered. Changes could include a readiness to increase output and adaptability into new product or service lines. As reported by ICT respondents in this survey, dealing with increased demand can represent a significant challenge as not only new clientele but the retention of existing customers is at stake. Steps that can be taken to ensure an organisation's ability to adapt and thrive include establishing on-going relationships, pre-event, with potential suppliers, collaborators and contractors; ensuring either adequate stocks or access to alternative suppliers; creating a workplace culture where employees will want to put the extra effort in for the organisation in a crisis setting; and ensuring that leaders are well informed and networked and able to recognise the opportunities that may arise.

10.6   POSTSCRIPT

Unfortunately, the September 2010 earthquake described in this report was a forerunner of a more devastating MM 6.3 shake which occurred on 22 February 2011. The epicentre of this earthquake was only 13 km south east of the Christchurch CBD and due to the unique geography of the area created ground accelerations four times higher than the 2011 Great East Japan MM 9 earthquake (GNS Science, 2011). This event resulted in 185 deaths and extensive damage to the CBD as well as many residential areas. The combined effect of these two major earthquakes, along with a further two aftershocks above magnitude 6, has resulted in the demolition of over 1000 commercial and 7000 residential dwellings (Canterbury Earthquake Recovery Authority, 2012). Residential losses are estimated at over NZD 12 billion, commercial at NZ 10–12 billion and infrastructure repairs to roads, bridges, fresh, waste and storm water systems at NZD 2–5 billion (Brownlee, 2012; Muir-Woods, 2012; Stronger Christchurch Infrastructure Rebuild Team [SCIRT], 2012). On-going research by the Resilient Organisations team considers the effects of these later events and further findings can be found at www.resorgs.org.nz.

REFERENCES

Alesch, D. J., Holly, J. N., Mittler, E. and Nagy, R. 2001. *Organizations at Risk: What Happens When Small Businesses and not-for-profits Encounter Natural Disasters*. Public Entity Risk Institute PERI, Fairfax, VA.

Benson, C. and Clay, E. J. 2003. *Disasters, Vulnerability, and the Global Economy*. Internet www.worldbank.org, E-mail feedback@worldbank.org, 1.

Benson, C. and Clay, E. J. 2004. *Understanding the Economic and Financial Impacts of Natural Disasters*. World Bank Publications, Washington, DC.

Boarnet, M. G. 1997. Business losses, transportation damage and the Northridge Earthquake. *Journal of Planning Literature*, 11, 476–486.

Brown, C., Seville, E., and Vargo, J. 2013. The role of insurance in organisational recovery following the 2010 and 2011 Canterbury earthquakes. *Resilient Organisations Research Report* 2013/04.

Brownlee, G. 2012. Budget 2012: Recovery of Canterbury on Track. Retrieved from http://feeds.beehive.govt.nz/release/budget-2012-recovery-canterbury-track.

Bruneau, M., Chang, S. E., Eguchi, R. T., Lee, G. C., O'Rourke, T. D., Reinhorn, A. M., Shinozuka, M., Tierney, K., Wallace, W. A. and von Winterfeldt, D. 2003. A framework to quantitatively assess and enhance the seismic resilience of communities. *Earthquake Spectra*, 19, 733.

Canterbury Earthquake Recovery Authority. 2012. *Earthquake Recovery Update Issue 9* (Vol. April). Christchurch, New Zealand Government.

Chang, S. E., Rose, A., Shinozuka, M., Svekla, W. D. and Tierney, K. J. 2001. Modeling earthquake impact on urban lifeline systems: Advances and integration, in Zhao, F.X. and Han, Z.J. (eds), *Earthquake Engineering Frontiers in the New Millennium*, Swets & Zeitlinger, The Netherlands.

Chang, S. E., Seligson, H. and Eguchi, R. T. 1996. Estimation of the economic impact of multiple lifeline disruption: Memphis light, gas and water division case study. *Technical Report NCEER*, 96.

Cochrane, H. 2004. Economic loss: Myth and measurement. *Disaster Prevention and Management*, 13(4), 290–296.

Committee on Assessing the Cost of Natural Disasters, Board on Natural Disasters, Commission on Geosciences, Environment, and Resources & National Research Council. 1999. The impacts of natural disasters: A framework for loss estimation, National Academy Press, Washington, DC.

Dietch, E. A. and Corey, C. M. 2011. Predicting long-term business recovery four years after Hurricane Katrina. *Management Research Review*, 34(3), 311–324.

Dillman, D. A. 1978. *Mail and Telephone Surveys*, Wiley, New York.

Eidinger, J., Tang, A. and O'Rourke, T. 2011. New Zealand Earthquake. *Report of the 4 September 2010 Mw 7.1 Canterbury*, Darfield, Reston, VA.

Finch, P. 2004. Supply chain risk management. *Supply Chain Management: An International Journal*, 9(2), 183–196.

Ganderton, P., Brookshire, D., McKee, M., Stewart, S. and Thurston, H. 2000. Buying insurance for disaster-type risks: Experimental evidence. *Journal of Risk and Uncertainty*, 20(3), 271–289.

Galbraith, C. S. and Stiles, C. H. 2006. Disasters and entrepreneurship: A short review. *Developmental Entrepreneurship: Adversity, Risk, and Isolation*, 5, 147–166.

Giovinazzi, S., Wilson, T., Davis, C., Bristow, D., Gallagher, M., Schofield, A. and Tang, A. 2011. Lifelines performance and management following the 22 February 2011 Christchurch Earthquake, New Zealand: Highlights of Resilience. *Bulletin of the New Zealand Society for Earthquake Engineering*, 44(4), 402–417.

GNS Science. 2011. Scientists find rare mix of factors exacerbated the Christchurch quake, http://www.gns.cri.nz/Home/News-and-Events/Media-Releases/Multiple-factors.

Gunasekaran, A., Rai, B. K. and Griffin, M. 2011. Resilience and competitiveness of small and medium size enterprises: An empirical research. *International Journal of Production Research*, 49(18), 5489–5509.

Kachali, H., Stevenson, J.R., Whitman, Z., Seville, E., Vargo, J. and Wilson, T. 2012. Organisational resilience and recovery for Canterbury Organisations after the 4 September 2010 earthquake. *Australasian Journal of Disaster and Trauma Studies*,1, 11–19.

Kunreuther, H. 2006. Disaster mitigation and insurance: Learning from Katrina. *The Annals of the American Academy of Political and Social Science*, 604(1), 208–227.

Lee, A., Seville, E. and Vargo, J. 2013. Developing a tool to measure and compare organisations' resilience. *Natural Hazards Review*, 14(1), 29–41.

McManus, S. 2008. *Organisational Resilience in New Zealand*. University of Canterbury, New Zealand.

Muir-Woods, R. 2012. 7. The Christchurch earthquakes of 2010 and 2011. *The Geneva Reports*, 93.

Munich Reinsurance Company. 1999. *A Year, a Century, and a Millennium of Natural Catastrophes are all Nearing their End – 1999 is Completely in Line with the Catastrophe Trend*. Munich Re review. Retrieved 30 May 2011, from http://munichre.com/.

Nigg, J. M. 1995. *Business Disruption Due to Earthquake-Induced Lifeline Interruption*. Preliminary Paper 220, Disaster Research Centre, University of Delaware.

New Zealand (NZ) Treasury Department. 2010. *Monthly Economic Indicators September 2010*. New Zealand Government.

Pelling, M. 2003. *Natural Disasters and Development in a Globalizing World*. Routledge, London.

RMBF, R. M. B. F. N. Z. 2011. Registered Master Builders says construction in recession again. Retrieved 30 May 2011, from http://www.masterbuilder.org.nz/index.asp?id=134.

Rose, A. and Lim, D. 2002. Business interruption losses from natural hazards: conceptual and methodological issues in the case of the Northridge earthquake. *Global Environmental Change Part B: Environmental Hazards*, 4(1), 1–14.

Seville, E., Brunsdon, D., Dantas, A., Le Masurier, J., Wilkinson, S. and Vargo, J. 2008. Organisational resilience: Researching the reality of New Zealand organisations. *Journal of Business Continuity and Emergency Planning*, 2(2), 258–266.

Souter, G. 2000. Risks from supply chain also demand attention. *Business Insurance*, 34(20), 26–28.

Statistics New Zealand. 2008. *New Zealand Business Demography Statistics (Structural): At February 2007*, New Zealand Government.

Statistics New Zealand. 2011. *Business Demography Statistics*. New Zealand Government, retrieved from stats.govt.nz.

Stronger Christchurch Infrastructure Rebuild Team (SCIRT). 2012. What we are doing. Retrieved from http://strongerchristchurch.govt.nz/about/what.

Tertiary Education Union (Producer). 2011, 19 May. Don't leave Christchurch to train trades on its own. Retrieved from http://teu.ac.nz/2011/04/dont-leave-christchurch-to-train-trades-on-its-own/.

Tierney, K. 1994. Business vulnerability and disruption: Data from the 1993 Midwest floods. Paper presented at the 41st North American Meetings of the Regional Science Association International, Niagara Falls, Ontario, November 16–20.

Tierney, K. and Dahlhamer, J. 1997. Business disruption, preparedness and recovery: Lessons from the Northridge earthquake. Preliminary Paper 257, Disaster Research Centre, University of Delaware.

Tierney, K. and Nigg, J. 1995. *Business Vulnerability to Disaster-Related Lifeline Disruption.* Preliminary paper 223, Disaster Research Centre, University of Delaware.

Tierney, K. and Webb, G. 2001. *Business Vulnerability to Earthquakes and Other Disasters.* Preliminary paper 320, Disaster Research Centre, University of Delaware.

TVNZ, T. N. Z. (Producer). 2011, 10 May. OCR cut could be much-needed boost to building industry. Retrieved from http://tvnz.co.nz/business-news/ocr-cut-could-much-needed-boost-building-industry-4052988.

USGS. 2009. Modified Mercalli intensity scale. Retrieved 1 June 2011, from http://earthquake.usgs.gov/learn/topics/mercalli.php.

Walker, B., Nilakant, V., van Heugten, K. and Rochford, K. 2013. Leading in a post-disaster setting: Guidance for human resource practitioners. *New Zealand Journal of Employment Relations*, 38, 1.

Vargo, J. and Seville, E. 2011. Crisis strategic planning for SMEs: Finding the silver lining. *International Journal of Production Research*, 49(18), 5619–5635. doi: 10.1080/00207543.2011.563902.

Webb, G., Tierney, K. and Dahlhamer, J. 1999. Predicting long-term business recovery from disaster: A comparison of the Loma Prieta Earthquake and Hurricane Andrew.

Whitman, Z., Kachali, H., Roger, D., Vargo, J. and Seville, E. 2013. Short-form version of the Benchmark Resilience Tool (BRT-53). *Measuring Business Excellence,* 17(3), 3–14.

Wood, J. S. 2008. The finance of Katrina. *International Journal of Social Economics*, 35(8), 579–589.

Index